AF564912

Basics of
NANO RESEARCH

Basics of
NANO RESEARCH

Sidharth Baliyan

ANMOL PUBLICATIONS PVT. LTD.
NEW DELHI-110 002 (INDIA)

ANMOL PUBLICATIONS PVT. LTD.
Regd. Office: 4360/4, Ansari Road, Daryaganj,
New Delhi-110002 (India)
Tel.: 23278000, 23261597, 23286875, 23255577
Fax: 91-11-23280289
Email: anmolpub@gmail.com
Visit us at: www.anmolpublications.com

Branch Office: No. 1015, Ist Main Road, BSK IIIrd Stage
IIIrd Phase, IIIrd Block, Bengaluru-560 085 (India)
Tel.: 080-41723429 • Fax: 080-26723604
Email: anmolpublicationsbangalore@gmail.com

Basics of Nano Research

First Edition, 2011

ISBN 978-81-261-4834-9

PRINTED IN INDIA

Printed at Suman Printers, Delhi

Contents

	Preface	*vii*
1.	Nano Framework	1
2.	Nano Processes	38
3.	Nano Research Data	80
4.	Nano Research Analysis	146
5.	Research Method	209
6.	Nano Research Report	243
	Bibliography	269
	Index	271

Preface

Advances in physics, molecular biology, and computer science are converging on the capacity to control, with molecular precision, the structure and function of matter. These twenty original contributions provide the first broad-based multidisciplinary definition and examination of the revolutionary new discipline of molecular engineering, or nanotechnology. They address the promise as well as the economic, environmental, and cultural challenges of this emerging atomic-scale technology.

Leaders in their field describe current technologies that feed into nanotechnology--atomic imaging and positioning, protein engineering, and the de novo design and synthesis of self-assembling molecular structures. They present development strategies for coordinating recent work in chemistry, biotechnology, and scanning-probe microscopy in order to successfully design and engineer molecular systems.

Author

1

Nano Framework

CONCEPTUAL STUDY

Various aspects of the marginality concept have been discussed, but the empirical findings are limited. There is a need for empirical data because of the importance of the topic and because of the benefits to the travel and tourism industry. This study further investigates the role of marginality factors (income, education, occupation and residence) in evaluating the travel behaviour of black and white Americans.

MARGINALITY

For the purpose of this study marginality is viewed as a state of lower socioeconomic status and those social conditions income, education, occupation and residence or the lack thereof. Low participation of Black Americans in travel activities and the travel behaviour differences between Black and White Americans can be attributed to limited economical resources. The underlying implication being that if Black Americans had (equal access to income, education, occupational status, and integrated housing,) access to the same travel and recreational activities, their participation patterns would be similar to White Americans.

RESEARCH QUESTION

The primary research question is: "Are there differences between black and white.

Americans with respect to travel behaviour due to marginality?"

This question is an extension of the work of Floyd as he did not address the influences of marginality. Floyd's study investigated the aspects of recreation rather than travel and tourism and discussed how individuals may be involved in recreational activities without traveling in vast distances to participate in these recreational activities. Additionally, two specific questions will be addressed.

The two questions that will be addressed are as follows:

- Is travel behaviour influenced by marginality?
- Is marginality effective in explaining the travel behaviour or differences in travel behaviour of black and white Americans?

RESEARCH OBJECTIVES

- To examine whether or not differences exist between black and white travellers with respect to selected travel behaviour variables.
- To examine the role (if any) that preferred activities participated in during leisure travel, types of trips selected, and length of stay have in understanding differences in travel behaviour of black and white travellers.
- To evaluate the extent to which blacks' and whites' socio-economic status is a predictor of their travel behaviour.

RESEARCH PROPOSITIONS

Based on the research objectives and the literature review, five specific research hypotheses and one hundred and twenty-three sub-hypotheses were formulated to guide the objectives of this study. Three propositions were also developed for this study.

The propositions of the study are:

- Black and White Americans leisure travellers differ in their length of stay, types of trips selected and activities participated in during leisure travel.
- Black and White American leisure travellers differ in income, education, occupation and residence as a predictor of their travel behaviour.

- Black and White American leisure travellers differ in their travel behaviour and these differences continue to exist when the populations are controlled for socio-economic status.

RESEARCH HYPOTHESES

- *Proposition* 1: Black and White American leisure travellers differ in their length of stay, type of trips selected and activities participated in during leisure travel.
- *H*1: Length of stay during leisure travel is a function of marginality.
- *H*2: Selecting trips is a function of marginality.
- *H*3: Travellers will participate in different activities during leisure travel as a function of marginality.
- *Proposition* 2: Black and White American leisure travellers differ in income, education, occupation and residence as a predictor of their travel behaviour.
- *Proposition* 3: Black and White Americans differ in their leisure travel behaviour and these differences continue to exist when the populations are controlled for socio- economic status (marginality).
- *H*4: Black travellers will differ from white travellers as a function of marginality.

DATA COLLECTION PROCEDURES

Population

The population for this study was a sample of black and white American travellers in six southeastern states in the U.S. so that a comparison of their travel behaviour and activities could be evaluated. Respondents were screened based on their self-described ethnic identification as this study was limited to black and white Americans only.

Participants were also screened on their travel pattern during the past year. To participate in this study, respondents were to have traveled away from home for least two nights during the past year.

Sampling Frame

The sampling frame consisted of extracted phone numbers from the survey sampling data base of New South Research in Birmingham, AL. Individuals who had taken at least one pleasure trip two nights away from home during the past year were qualified to participate in this study. For the purpose of this study, participants residing in the country only were surveyed.

Participants were selected from Alabama, Georgia, Mississippi, Louisiana and South Carolina and Tennessee. To achieve the sample, the systematic sampling method was used. This strategy followed random start, every Kth sample unit was selected from the population.

Telephone Surveys

For the purpose of this study a professional telephone survey will be used. A professional telephone survey is a telephone survey done with personal supervision.

Advantages of Telephone Surveys

Unlike personal interview surveys and mail surveys, telephone surveys have several advantages. Telephone surveys offer sampling advantages such as higher completion rates, greater level of cooperation, and usefulness in approaching hard to interview populations.

As it relates to subject matter in telephone surveys, the advantages are that embarrassing topics can be covered, controversial topics can be covered and immediate topics can be addressed.

In the interviewing process several advantages include a high quality of interviewing, a decrease in interview bias and there is no third party influence.

Limitations of Telephone Surveys

A bias towards households that do not have a telephone represents a limitation of the telephone survey method. A second potential bias is the problem in attaining households with unlisted telephone numbers.

Although random digit dialing will minimize or almost eliminate this problem, it still should be mentioned as a possible limitation of the telephone survey method. For the purpose of this study telephone numbers will be provided by the inquirers.

Interviewer Selection

Trained interviewers of New South Research are selected on the basis of their telephone voice, dependability, trainability, possible predictability and the ability to record well. General training will be done by New South Research. In-house instructions will include conducting the interview, how to handle probes, and the handling of other specifics such as telephones lines, location of calls, etc.

Practice interviews will include actual telephone calls to real respondents. Interviewers will be required to follow a specified script which introduces them and explains their reason for calling. Interviewers will also be required to keep a record of their dialing activities by completing a dialing sheet.

All interviewing was conducted from a central telephone facility in Birmingham, Alabama. Prior to the beginning of interviewing, the interviewers were thoroughly briefed on the purposes of the survey and each question was reviewed in detail with the interviewing staff to ensure they clearly understood the question being asked and the possible answers they might receive. While interviewing were being conducted, a supervisor was present at all times to answer questions cr resolve problems that might arise.

During the course of this study, no problems were encountered which would affect the quality of the data or the research results. As surveys were completed, each interview was carefully checked by the supervisor for completeness and consistency."

Confidence Limits

Based on a total sample of this size (500), it is expected that the results obtained from this research will be within +5.7

per cent of that which would have been obtained if all households in the area had been interviewed. Further, it is expected that will occur 95 times out of 100.

SURVEY INSTRUMENT

The survey instrument was developed to investigate factors influential in travel behaviour among black and white Americans.

The questionnaire contains two sections designed to attain the required information for the purpose of the study. Section one consists of questions designed to gather information on the length of stay, information sources, types of trips selected, and preferred activities (activities participated in during leisure travel).

The second section consists of questions to collect demographic data such as age, gender, marital status, income, education, occupational level, religious patterns, residence and race/ethnic group. Once the telephone survey was completed it was numerically coded, and statistical analyses were used to test the null hypotheses utilizing empirical data.

NON-RESPONSE BIAS

Non-response bias was checked by a random sample of 25 individuals who were not included in the telephone survey to compare the characteristics of the respondents and determine if non-respondents are significantly different from the final respondents of the study.

Non-respondents were asked several questions from the survey. Demographic information such as race, gender, age, income, education and residence was be collected. The results of the analysis on non-respondents will be discussed in chapter four.

VARIABLES

Race

For the purpose of this study only two categories of race were included. Race was assessed by a self-identification question. In other words how one identifies himself and responded to question 2 in the survey determined their inclusion in this study.

Marginality

For the purpose of this study, marginality was determined by the level of income, education, occupation and place of residence. Respondents were asked to indicate their range of household income, highest education level completed, type of occupation and where they resided. The indicators income, education, occupation and residence were selected in that they are important in determining accessibility and level or presence of discrimination.

Types of Trips Selected

An examination of the types of trips respondents chose to select led to the development of a list of eight trip types. Respondents were asked to select the type of your most recent pleasure trip.

Length of Stay

The amount of time spent on each trip was acquired. Respondents were asked to indicate how long they stayed on their most recent pleasure trip.

Activities Participated In

Activities travellers actually participated in during leisure travel were measured by an open-ended question. In question 5 respondents were asked, Please indicate the activities you participated in during your most recent pleasure trip." Possible responses to this question were grouped into fourteen categories. These categories are based on Kaplan's Taxonomy, 1960 and factor analysis updates by Noe(1974) and McLoughlin and Noe.The activity categories include immobile activities, sports, exercise-health. popular art, Association-Sociability, outdoor-individual, hunting and fishing, games, fine art, camping-hiking, mobile activities, golf, risk-skill and boating-skiing It is understood that respondents might have participated in several activities during their leisure travel.

DATA ANALYSIS

To effectively complete an analysis of the data, a quantitative method of analysis was applied. The hypotheses corresponding to the study objectives were analysed using the SAS (Statistical Analysis Software) program. New South Research used Dbase 4 to analyse their data. All results were considered to be statistically significant at the.05 or better probability level.

This study assumes that the individuals who responded to this survey were truthful. Finally the study assumes that the survey instrument used to collect data actually measures the variables which determine the travel behaviour of black and white leisure travellers.

All research is subject to certain limitations. In this case, since a telephone survey was conducted, the results do not include opinions for certain populations. Included among these populations are persons who do not have a telephone in their home, those living in institutional quarters and those living with someone else such as parents living with their children.

New listings and persons with unlisted numbers included in the research and were reached by random digit "1-plus dialing. That is, "1" was added to each working number to reach these sub-segments of the population.

While the attitudes of those persons excluded from the survey may or may not be different from those who were included in the research, the omission of these segments is not expected to have a significant effect on the survey results because of the small size of the segment."

RESEARCH METHODOLOGY

INTRODUCTION

Research is the cornerstone of any science, including both the hard sciences such as chemistry or physics and the social (or soft) sciences such as psychology, management, or education.

It refers to the organized, structured, and purposeful

attempt to gain knowledge about a suspected relationship. Many argue that the structured attempt at gaining knowledge dates back to Aristotle and his identification of deductive reasoning. Deductive reasoning refers to a structured approach utilizing an accepted premise (known as a major premise), a related minor premise, and an obvious conclusion.

This way of gaining knowledge has been called a syllogism, and by following downward from the general to the specific, knowledge can be gained about a particular relationship.

An example of an Aristotelian syllogism might be:

- *Major Premise*: All students attend school regularly
- *Minor Premise*: John is a student
- *Conclusion*: John attends school regularly

In the early 1600s, Francis Bacon identified a different approach to gaining knowledge. Rather than moving from the general to the specific, Bacon looked at the gathering of specific information in order to make general conclusions.

This type of reasoning is called inductive and unlike Aristotelian logic allows new major premises to be determined. Inductive reasoning has been adopted into the sciences as the preferred way to explore new relationships because it allows us to use accepted knowledge as a means to gain new knowledge.

For example:

- *Specific Premises*: John, Sally, Lenny and Sue attended class regularly
- *Specific Premises*: John, Sally, Lenny, and Sue received high grades
- *Conclusion*: Attending class regularly results in high grades

Researchers combine the powers of deductive and inductive reasoning into what is referred to now as the scientific method.

It involves the determination of a major premise (called a theory or a hypothesis) and then the analysis of the specific examples (research) that would logically follow. The results might look something like.

Major Premise:	Attending classes regularly results in high grades	
Class Attendance: (Suspected Cause)	Group 1:	John, Sally, Lenny and Sue attend classes regularly
	Group 2:	Heather, Lucinda, Ling, and Bob do not attend classes regularly
Grades: (Suspected Effect)	Group 1:	John, Sally Lenny, and Sue received A's and B's
	Group 2:	Heather, Lucinda, Ling, and Bob received C's and D's
Conclusion:	Attending class regularly results in higher grades when compared with not attending class regularly (the Major Premise or Hypothesis is therefore supported)	

Utilizing the scientific method for gaining new information and testing the validity of a major premise, John Dewey suggested a series of logical steps to follow when attempting to support a theory or hypothesis with actual data. In other words, he proposed using deductive reasoning to develop a theory followed by inductive reasoning to support it. These steps can be found in Table.

Dewey's Scientific Method:

- Identify and define the problem
- Determine the hypothesis or reason why the problem exists
- Collect and analyse data
- Formulate conclusions
- Apply conclusion to the original hypothesis

The steps involved in the research process can vary depending on the type of research being done and the hypothesis being tested.

The most stringent types of research, such as experimental methods (sometimes called laboratory research), contain the most structured process. Naturalistic observation, surveys, and other non-intrusive studies, are often less structured. A general process guide for doing research, especially laboratory research are given below.

Steps Involved in the Research Rocess:

- Determine your theory or educated quess about a relationship. Involves identifying and defining a problem and reviewing the current literature.
- Operationally define all variables to be involved in the research.
- Develop hypothesis by plugging variables into original theory. A hypothesis is a testable theory with operationally defined variables.
- Standardize the research methods by developing a research protocol to be used with every subject. Include in this step the methods for subject selection and assignment as well as how you will attempt to control for any extaneous variable.
- Select subjects and assign to groups using the protocol developed in step 4.
- Test subjects using the protocol developed in step 4.
- Analyse results.
- Determine the significance of the results and how these results compare with other studies. Critique your research and suggest needs for further research based on your findings.
- Communication results through journal publication, book, book chapter, report, presentation, or any means that will benifit those to whom the research applies.
- Replicate. While part of the research process, this step is most often completed by a different researcher.

DETERMINING A THEORY

While you may see a theory as an absolute, such as the theory of gravity or the theory of relativity, it is actually a changing phenomenon, especially in the soft or social sciences. Theories are developed based on what is observed or experienced, often times in the real world. In other words, a theory may have no additional backing other than an educated guess or a hunch about a relationship. For example, while teaching a college course in research, I notice that nontraditional students tend to be more involved in class lectures and perform better on class exams than traditional students. My theory, then, could be that older students are more dedicated to their education than younger students.

At this point, however, I have noticed only a trend within a single class that may or may not exist. I have developed a theory based on my observations and this theory, at least at this point, has no practical applications. Most theories are less concerned with application and more concerned with explanations.

For example, I could assume, based on my observations, that older students have witnessed the importance of education through their work and interactions with others. With this explanation, I now have a theoretical cause and effect relationship: Students who have had prior experience in the workforce are more dedicated to their education than students who have not had this experience.

Before moving beyond this point it is always wise to do a literature review on your topic and areas related to your topic. Results from this search will likely help you determine how to proceed with your research. If, for example, you find that several studies have already been completed on this topic with similar results, doing yet another experiment may add little to what is already known. If this is the case, you would need to rethink your ideas and perhaps replicate the previous research using a different type of subject or a different situation or you may choose to scrap the study all together.

DEFINING VARIABLES

Variables can be defined as any aspect of a theory that can vary or change as part of the interaction within the theory. In other words, variables are anything can effect or change the results of a study. Every study has variables as these are needed in order to understand differences.

In our theory, we have proposed that students exposed to the workforce take a more active role in their education than those who have no exposure. Looking at this theory, you might see that several obvious variables are at play, including 'prior work experience' and 'age of student.' However, other variables may also play a role in or influence what we observed. It is possible that older students have better social skills causing them to interact more in the classroom.

They may have learned better studying skills, resulting in higher examination grades. They may feel awkward in a classroom of younger students or doubt their ability more and therefore try harder to succeed. All of these potential explanations or variables need to be addressed for the results of research to be valid. Let's start with the variables that are directly related to the theory. First, the prior work experience is what we are saying has the effect on the classroom performance. We could say that work history is therefore the cause and classroom grades are the effect. In this example, our independent variable (IV), the variable we start with (the input variable) is work experience. Our dependent variable (DV), or the variable we end up with (the outcome variable) is grades.

We could add additional variables to our list to create more complex research. If we also looked at the affect of study skills on grades, study skills would become a second independent variable. If we wanted to measure the length of time to graduation along with grades, this would become a second dependent variable. There is no limit to the number of variables that can be measured, although the more variables, the more complex the study and the more complex the statistical analysis.

The most powerful benefit of increasing our variables, however, is control. If we suspect something might impact our outcome, we need to either include it as a variable or hold it constant between all groups. If we find a variable that we did not include or hold constant to have an impact on our outcome, the study is said to be confounded. Variables that can confound our results, called confounding variables, are categorized into two groups: extraneous and intervening.

Extraneous Variables

Extraneous variables can be defined as any variable other

than the independent variable that could cause a change in the dependent variable. In our study we might realise that age could play a role in our outcome, as could family history, education of parents or partner, interest in the class topic, or even time of day, preference for the instructor's teaching style or personality. The list, unfortunately, could be quite long and must be dealt with in order to increase the probability of reaching valid and reliable results.

Intervening Variables

Intervening variables, like extraneous variables, can alter the results of our research. These variables, however, are much more difficult to control for. Intervening variables include motivation, tiredness, boredom, and any other factor that arises during the course of research. For example, if one group becomes bored with their role in the research more so than the other group, the results may have less to do with our independent variable, and more to do with the boredom of our subjects.

DEVELOPING THE HYPOTHESIS

The hypothesis is directly related to a theory but contains operationally defined variables and is in testable form. Hypotheses allow us to determine, through research, if our theory is correct. In other words, does prior work experience result in better grades? When doing research, we are typically looking for some type of difference or change between two or more groups. In our study, we are testing the difference between having work experience and not having work experience on college grades. Every study has two hypotheses; one stated as a difference between groups and one stated as no difference between groups.

When stated as a difference between groups, our hypothesis would be, "students with prior work experience earn higher grades than students without prior work experience." This is called our research or scientific hypothesis. Because most statistics test for no difference, however, we must also have a null hypothesis. The null hypothesis is always written with the assumption that the groups do not differ. In this study, our null hypothesis would state that, "students with work experience will not receive different grades than students with no work experience."

The null hypothesis is what we test through the use of statistics and is abbreviated H_0. Since we are testing the null, we can assume then that if the null is not true then some alternative to the null must be true. The research hypothesis stated earlier becomes our alternative, abbreviated H_1. In order to make research as specific as possible we typically look for one of two outcomes, either the null or the alternative hypothesis.

To conclude that there is no difference between the two groups means we are accepting our null hypothesis. If we, however, show that the null is not true then we must reject it and therefore conclude that the alternative hypothesis must be true. While there may be a lot of gray area in the research itself, the results must always be stated in black and white.

STANDARDIZATION

Standardization refers to methods used in gathering and treating subjects for a specific study. In order to compare the results of one group to the results of a second group, we must assure that each group receives the same opportunities to succeed. Standardized tests, for instance, painstakingly assure that each student receives the same questions in the same order and is given the same amount of time, the same resources, and the same type of testing environment. Without standardization, we could never adequately compare groups.

For example, imagine that one group of students was given a particular test and allowed four hours to complete it in a quiet and well lit room. A second group was given the same test but only allowed 30 minutes to complete it while sitting in a busy school lunchroom full of laughing and talking children. If group 1 scored higher than group 2 could we truly say that they did better? The answer is obviously 'no.' To make sure we can compare results, we must make everything equal between the two or more groups. Only then could we say that group 1 performed better than group 2. Standardization of the research methods is often a lengthy process. The same directions must be read to each student, the same questions must be given, and the same amount of time must be assured. All of these factors must be decided before the first subject can be tested. While standardization refers mainly to the testing situation itself, these principles of 'sameness' involve the selection of subjects as well.

VARIABLES AND CONSTANTS

Much of the scientific study of psychology involves the investigation of what are called *variables*. Unfortunately, variables are a bit difficult to define, without using the word *variable*. One way to approach this is to contrast a variable with its opposite, a *constant*. For our purposes, a constant is some characteristic that does not vary among the members of a group we wish to study. A variable is a characteristic that does vary among members of that group. That characteristic can be a behaviour or some psychological characteristic such as intelligence, friendliness or preference for music.

A situation may also be a variable because some people may be in a situation that others are not. I do not like dictionary definitions but if I were to attempt one for the word *variable*, it would be something like: *a behaviour, psychological characteristic, or situation that is not the same for all members of a group who are being studied.* Among members of a PSY 101 class we might be interested in investigating Grade Point Average. GPA is a variable because it is a characteristic that varies among members of the class.

A typical question a psychologist might ask is why some students have higher GPAs than others. To answer that question, the psychologist would probably measure GPA in some way, measure the levels of one or more other variables, and compare the outcomes. It should be easy for you to imagine some other variables that might have some relationship to GPA. Some examples are: time spent studying, time spent working at outside jobs, chronological age, class level, high school GPA. Any of these, and many more that might occur to you, could be part of the explanation for GPA differences within members of a class. All of these are variables and any of them could be investigated. When you get to class I will ask you for some more variables —however far-fetched—that might have a relationship to GPA.

In contrast, no useful information would result from investigating the effect of a constant upon GPA. For example, all members of the class are breathing, so an investigation of the effect of breathing or not upon GPA would be useless. All members of the class are wearing clothes so there would be no way to use this group to investigate the effects of nakedness upon GPA. If this discussion of constants seems strange, you should remember that the primary reason for introducing the concept of the constant here is to help to define *variable.*

REPLICATION

In scientific accounts of research, descriptions of variables must be sufficiently specific so that another researcher could re-do the study using the same variables. It is very basic to science that the events studied by one researcher could also be studied by another scientist in another time and place. This is called *replication*. Confirmation of findings through replication is an important way in which the power, or certainty, of scientific findings can be increased.

Specificity of Variables

If you look at the variables that might help to explain GPA listed above, you will notice that they are quite specific. Science can only deal with specific, observable events. Even though the examples given above are quite specific, they could be even more so.

For example, take the variable "time spent working at outside jobs." Try to imagine some ways in which that variable could be made so specific that two people in different colleges could measure it in PSY 101 classes and have confidence that the same thing is being measured. Does volunteer work count? This variable could be made more specific by saying "time spent working at outside jobs for money". Alternatively, the variable could be restated "time spent working at outside jobs whether volunteering or being paid." Either way, the variable is more likely to be measuring the same thing if it were to be used by different researchers.

MEASUREMENT OF VARIABLES

Another issue surrounding the definitions of variables is the question of measurement. Even if the variable itself is defined quite specifically, it is also important to be specific about the way in which it will be measured. For example, in the investigation of GPA and other variables, it is important to know how GPA will be measured. The students could be asked to self-report their GPA to two decimal places. Alternatively they could be asked to self-report it within ranges involving no decimals, *i.e.* 0 to 1, 1 to 2, 2 to 3 and 3 to 4. In some circumstances it might be possible to get lists of GPAs from the college registrar.

These three tactics would almost certainly yield different results and so it is important to be specific about the measurement of the variable if we want to be sure that studies of GPA at two different colleges were measuring the same thing. Measurement of GPA is easy compared to the other variables, listed above, that might be studied for their effect upon GPA. Look at some of them and come to class prepared to suggest ways in which they could be measured.

OPERATIONAL DEFINITIONS

In summary, scientific research reports define variables specifically and in terms of measurement. These definitions are called *operational definitions,* because the variable is defined in terms of the operations needed to measure it. The primary goal of this course is to teach you to critically evaluate assertions about behaviour and one of the most basic ways to be critical is to ask about the operational definitions of variables. Operational definitions are particularly important when discussing difficult-to-observe psychological characteristics such as being depressed, being smart or being sensitive. Someone might say, for example, that a new study had shown that first-born children are more likely to be withdrawn whereas second-born children are more likely to be outgoing.

When you hear something like that you can begin to evaluate the assertion by asking for the operational definitions. When you come to class, I am going to ask you to make some suggestions about how these things might be operationally defined and I will also ask you to consider potential pitfalls in the operational definitions that might have been used in this study of birth order and personality.

One reason why an understanding of operational definitions is such an important critical tool is that operational definitions can vary considerably in quality. If you were doing a study of depression and air temperature, a thermometer might provide a good operational definition of air temperature. Alternatively, you could open the window and stick your head out and make a guess based on this experience. During this course we will often evaluate the adequacy of operational definitions.

COMMON RESEARCH METHODS

There are many different types of research methods, also called research designs, that are used by psychologists in trying to find things out about behaviour. This is just a quick aid to the identification of research designs. In real life, some studies may combine the features of several research designs or may contain elements not included below.

Experiment

Participants randomly assigned to different groups being studied. Groups are treated differently in one or a few very specific ways—the independent variable. Behaviour resulting from this treatment difference is measured—the dependent variable. If one group gets a specific treatment and ones does not, usually the treated group is called the experimental group and other groups are called control groups. Conditions other than the independent variable are held as constant as possible for all groups. These constant conditions are called controls. If participants are their own control group, that is, they receive both research treatments, the design is called a within-subjects experiment. Conclusions can be taken to indicate a cause and effect relationship between the independent and dependent variables. Because of this, the experiment is in a class by itself and it is a very special type of research procedure.

Quasi-experiment

Participants achieve membership in different groups as a result of characteristics other than random assignment, for example: gender, age, socioeconomic status, athletic ability, or ethnic identification. A link may be found between one or more of these characteristics and some outcome variables, but cause and effect relationships are not clearly identified. Without random assignment to groups, a researcher cannot clearly demonstrate cause.

Correlational Study

In the most general sense, a correlational study investigates the relationship between two variables. Usually the data are reported as correlation coefficients. Strength and direction (positive or negative) of relationships can be demonstrated by correlational studies but causal links remain an open question.

Longitudinal Study

A longitudinal study follows a group composed of the same people across a period of the life span. The behaviour of these individuals is observed and/or measured at several intervals over time in an attempt to study the changes in their behaviour. Longitudinal studies may cover a short time, such as a few weeks, or a long time, such as the entire life span. Longitudinal studies may additionally employ other methods, such as quasi-experimental or correlational approaches, but the defining characteristic is that the same people are studied repeatedly across time.

Cross Sectional Study

A cross sectional study usually examines groups of different people who belong to different age groups as a means of studying behaviour development across part or all of the life span. These studies can usually be done more easily and quickly than longitudinal studies but the resulting data may be of lower quality. More rarely, the term cross sectional may be used to describe studies which divide and examine segments of society based on variables other than age, such as income, educational level or family size.

Survey

A survey is a structured list of questions presented to people. Surveys may be written or oral, face to face or over the phone. It is possible to cheaply survey large numbers of people, but the data quality may be lower than some other methods because people do not always answer questions accurately.

Interview

An interview may be highly structured or it may involve less structured narrative. It may include survey methodology. It usually involves people responding orally to questions or talking about their thoughts on a topic.

Case study

A case study involves extensive observations of a few individuals. Data collection may include watching behaviour, interviews and record searching. Case studies may be retrospective and/or prospective. Usually case studies are employed where the behaviour or situation is so rare that other methods, involving larger groups of participants, are not possible.

Naturalistic Observation

Naturalistic observations can range from unstructured observations of humans or other animals to situations involving hypothesis testing or some manipulations of a natural setting. If you wanted to know if males are likely to hold doors open for females, you could watch until you had seen a number of natural occurrences of this, or you could get a female helper to follow males into buildings and watch to see what happens. It can be difficult to precisely define the natural setting, particularly when the participants are humans. Placing an actual research procedure into this category or others can involve a judgment call which might be debatable.

Demonstration

An unsystematically engineered observation of behaviour, sometimes involving only one participant. The demonstration is remarkably common in the history of psychology, even though it provides only very weak evidence. It is not a recognized research method but it is a term which can be quite useful as a descriptor for studies that seem to employ no established method.

FORMULATING THE RESEARCH PROBLEM

Researchers organize their research by formulating and defining a research problem. This helps them focus the research process so that they can draw conclusions reflecting the real world in the best possible way.

HYPOTHESIS

In research, a hypothesis is a suggested explanation of a phenomenon. A null hypothesis is a hypothesis which a researcher tries to disprove. Normally, the null hypothesis represents the current view/explanation of an aspect of the world that the researcher wants to challenge.

Research methodology involves the researcher providing an alternative hypothesis, a research hypothesis, as an alternate way to explain the phenomenon.

The researcher tests the hypothesis to disprove the null hypothesis, not because he/she loves the research hypothesis, but because it would mean coming closer to finding an answer to a specific problem. The research hypothesis is often based on observations that evoke suspicion that the null hypothesis is not always correct.

The researcher tests the hypothesis to disprove the null hypothesis, not because he/she loves the research hypothesis, but because it would mean coming closer to finding a answer to a specific problem. The research hypothesis is often based on observations that evoke suspicion that the null hypothesis is not always correct.

In the Stanley Milgram Experiment, the a null hypothesis was that the personality determined whether a person would hurt another person, while the research hypothesis was that the role, instructions and orders were much more important in determining whether people would hurt others.

VARIOUS FUNCTIONS OF MANAGEMENT

INTRODUCTION

Mugging up management principles can't learn management. Management is best learned through handling it systematically. A practical exposure of management principles is the foundation of the career development of any management student. When the theories are actually applied in the market, many external environments factors affect it, thereby, changing its dimensions. Theories provide the fundamental stone for guidance of practice examines the element of the truth lying in the theory.

DEFINING MARKETING RESEARCH

- Marketing Research is a systematic and objective study of problems pertaining to the marketing of goods and services. It is not restricted to any particular area of marketing but is applicable to all its phases and aspects.
- The traditional definition of Marketing Research by American Marketing Association (AMA) -the systematic gathering, recording and analyzing of data about problems relating to the marketing of goods and services.-

- Also defined as the function which links the consumer, customer and public to the marketer through information used to identify and define marketing opportunities and problems. It involves the use of surveys, tests, and statistical studies to analyse consumer trends and to forecast the size and location of markets for specific products or services.

APPLICATION KNOWLEDGE

Marketing research revolves around the study of different brands of television. Profits flow back the cost and sales are the basis of profits. So this Marketing Research is carried on to get a feedback from the dealers about different characteristics like- market size, market share, price and others alike of various brands sold in certain locations. This research gives us an opportunity to apply our conceptual skills in practice and to learn the art of conducting study and presenting its findings in a systematic and scientific way.

FUNCTIONS OF MARKETING RESEARCH

- To provide a point at which all the buying of research is made, so that there is one point to manage it to reduce overlap and duplication.
- To provide a centre of expertise for the management of the purchasing of research.
- To manage the supplier-buyer interface.
- To ensure that the research purchased has a design appropriate for the job
- To manage the research process from inception to results and further.
- To help with the integration of different information sources to a common end.
- To act as a custodian of the marketing knowledge and history of the company's brands.
- To provide a single place where market research reports are kept and where access to them is available to the whole organization.
- To help with the creation of an information climate

for the company with respect to the markets in which it trades.

INFLUENCING FACTORS IN MARKETING RESEARCH

Research into all factors mainly social, cultural, psychological and personal provides marketers with clues to reach and serve consumers more effectively.

The starting point for understanding Marketing Research is consumer behaviour. Marketing and environmental stimuli enter the consumer's consciousness. Sets of psychological processes combine with certain consumer characteristics to result in decision processes and purchase decisions. The marketer's task is to understand what happens in consumer's consciousness between the arrival of the outside marketing stimuli and the ultimate purchase decisions.

Four key psychological processes-motivation, perception, learning and memory- fundamentally influence consumer responses to the various marketing stimuli. To understand how consumer actually takes the buying decisions, marketers must identify who makes and has input into the buying decision. People can be

- Initiators
- Influencers
- Deciders
- Buyers or Users

Different marketing campaign might be targeted to each type of person.

MARKET VS MARKETING RESEARCH

'Market' research is simply research into a specific market. It is a very narrow concept. 'Marketing' research is much broader. It not only includes 'market' research, but also areas such as research into new products, or modes of distribution such as via the Internet.

CONCLUSION

A research is a step-by-step process. It is very difficult

for any researcher to master all these steps without proper guidance. Application of management principles in all branches whether marketing, sales, personnel, finance and others alike in more effective utilization of available resources enlightens us with deeper knowledge of marketing.

TYPES OF RESEARCH

FOR WEB ENTREPRENEURS

Mastering which types of research methodology to use, and when, can make or break your Web business. Before I reveal which types are the best, let me describe what you have to research, as a Web entrepreneur.

You will have to do fairly intensive and exhaustive research during the planning phase of your Web business. This activity will have far reaching consequences, because the results will become the very foundation of your entreprise.

You will (should) also continue doing research during normal business operations, thereafter. This activity will not be as demanding, but will nevertheless be an important factor in your ability to stay ahead of the competition. Your research, if done properly, will help you "own" your niche market.

Therefore, it is imperative that you master the types of research methodology that will best serve your purpose, at each stage of your business development.

SEARCHING FOR ANSWERS

Aspiring Web entrepreneurs have to find answers to a number of critical questions, right from the very start. Here are some of them, and their respective types of research methodology.

- What is the demand like?
- Are there enough "profitable" keywords in the chosen domain to make the business venture worthwhile?
- What is the demand, and the supply for each keyword?
- How many keywords per page is best?

- How many unique visitors do you need before monetizing?
- How many Web pages of useful content do you need before submitting your Web site to directories?
  What will be your main business activity?
- If you have many business ideas, how will you recognize the one that will most likely be profitable?
- What will be the one for which you are best suited?
- If you do not already have a specific business idea in mind, how will you find out if you have what it takes to have a Web business?
- You should not come to the hasty conclusion that a Web business is not for you! You should find out for certain through proper research.
- Is the desired domain name available?
- What are the possible derivatives?What would the best domain name be for my Web business?
- How much are advertisers paying for the keywords in your chosen domain?
- What is the competition like?
- How many monetizing possibilities will you have?
- How many affiliate programs would be suitable?
  What types of monetizing are most suitable for my business?

METHODS

Web entrepreneurs use types of research methodology specifically designed for the Web environment. These methods require specifically designed software tools, obviously. Well.it may not be obvious to you yet, but trust me, it will be if you investigate.

TYPES OF RESEARCH METHODS

Basic Research

Answer fundamental questions about the nature of behaviour. Not done for application, but rather to gain knowledge for sake of knowledge.

For Example, look at the titles of these publications:

- Short and long-term memory retrieval: A comparison of effects of information overload and relatedness.
- Electrophysiological activity in the central nucleus of the amygdala: Emotionality and stress ulcers in rats.

Some people erroneously believe that basic research is useless. In reality, basic research is the foundation upon which others can develop applications and solutions. So while basic research may not appear to be helpful in *the real world*, it can direct us towards practical applications such as, but definitely not limited to:

- Skinner-trained animals to work for reinforcement-lead to work schedules and applications in I/O psychology, therapy, and education.
- All those therapeutic techniques that clinical psychologists and other therapists use to help people must studied to determine which are most effective for which situations, people, and problems.

Applied Research

Concerned with finding solutions to practical problems and putting these solutions to work in order to help others.

Some examples of publication titles:

- Effects of exercise, relaxation, and management skills training on physiological stress indicators.
- Promoting automobile safety belt use by young children.

Today, there is a push to more applied research. This is no small part due to the perspective in the United States where we want solutions and we want them now! BUT, we still need to keep our perspective on the need for basic research.

Program Evaluation

look at existing programs in such areas as government, education, criminal justice, etc., and determine the effectiveness of these programs.

Does thr program work?

For example-Does capital punishment work? Think of all the issues surrounding this program and how hard it is to examine its effectiveness. The most immediate issue, how do you define the purpose and "effectiveness" of capital punishment? If the purpose is to prevent convicted criminals from ever committing that same crime or any other crime, than capital punishment is an absolute-100% effective. However, if the point of capital punishment is to deter would-be criminals from committing crimes, then it is a completely different story.

BUSINESS PROBLEMS BY RESEARCH

PROBLEMS IN ESTABLISHING BUSINESS

The process of starting or launching business involves two aspects like commercial aspect and the legal aspect. The former aspect is decided by the businessman himself. While deciding the commercial aspect he depends on the legal aspects. The problems of starting a business enterprise involves what is to be done and how it is to be done. The following problems are encountered while starting a business enterprise:

Determination of the Objective

The basic problem behind the establishment of a new enterprise is the determination of the main objective. The objectives set for the organization should be clearly described and it should not be unrealistic. Besides setting up the basic objectives, the determination of subsidiary long-run objectives and short-run objectives also create a number of problems.

Development of an Idea

The entrepreneur develops the idea of starting business opportunities. He may visualize the business idea from himself or borrow from some experts. Selection of business has been regarded as one of the significant problem at the time of starting new business.

Product Analysis and Market Research

While selecting the nature and the type of business, one

has to decide the types of product be produced or purchased and the type of market needed. The determination of the type of product and market, it require product and market analysis. The product so selected should be attractive in shape, size and colour so that it can satisfy customers at large.

Selection of Forms of Organization

After selection of the product and the type of market, the next problem is the selection of the forms of organization. There are various forms of organization like sole trader, partnership, company and co-operative form of organization. The selection of suitable form of organization depends on nature and size of business, the amount of finance available and the managerial needs.

Determination of Scale of Operation

The next problem at the time of establishing a business unit is the determination of the scale of operation. Whether business will be carried on large scale or small scale, an experienced person with sufficient capital and hoping certainly of income may enter into large scale production.

Estimating Capital Requirement

Capital is regarded as the life-blood of the business unit. Before starting the business, it's capital requirement should be properly decided. The requirement of capital is determined on the basis of requirement for fixed capital, working capital is preliminary expenses. The amount of capital on the above directions is to be determined before starting the business enterprise.

Selection of Business Site

Selection of site include plant location, plant building and equipment and lay-out-for the plant. These factors are essential for the smooth conducting of the business.

Selection of Staff and other Employees

The selection of right man in the right job is an important

factor which facilitates smooth functioning of the organization. Selection of right staff and employee is a problem at the time of starting a new business.

Completion of Legal Formalities

The businessman has to complete all the legal formalities at the time of starting a new business enterprise. These formalities differ organisation to organisation.

It is not difficult to get your hydroponic system set up. Most of the Hydroponic Gardening guides for beginners include a section on build-it-yourself hydroponic system. They provide a parts list, a tools list, and simple step-by-step instruction on how to build your own hydroponic system.

However, once the hydroponic unit is in operation, quite often, the beginners will discover problems, some of which cannot be easily solved after the system has already been built. The followings are some of the common problems encountered.

Problem#1

There is a concern on just how much nutrients to be poured over the aggregate. Because for those hydroponic systems using a "light proof" container concept, you will not be able to see through the containers or down through the aggregate. So it is very difficult to gauge the amount or level of nutrient solution. Without this visibility, it is quite likely that the plants will be killed by either under or overfilling.

The viable solutions can be either put a visual indicator showing the nutrient solution level or water sensors for automatic system.

Problem#2

The second problem is how often to pour nutrient over the aggregate. If you just follow the interval for your house plants, for an example, three to five times a week, you would probably kill your plants. For hydroponics, because of the wider air gap in the aggregate as compared to soil, the nutrient solution will tend to evaporate from the aggregate much more quickly than water from soil. So in general, you would need to supply nutrient to your plants at least once a day. The more simple the system, the more frequently you will have to be around to add nutrient solution. The interval can be anywhere from one to four times a day depending on several factors, such as light, temperature, humidity, type and size of your plants, and the size of your container.

Problem#3

A third problem involves proper aeration (or supply of air or oxygen) for the plants' roots. This area usually is not a concern for soil gardening in the backyard because worms perform this function. In hydroponic system, particularly those using PVC pipes with holes drilled for plants, too often the roots clog up the waterways and aeration in the root zone may become a problem.

Different systems will have different ways of providing proper aeration, for examples, using pumps, raised platforms or specific aggregate suitable for hydroponics.

Simple Problems

To some, these problems seem to be a matter of common sense. However, if you are new to this soil-less gardening concept and without going through the actual exercise once, you are likely to discover a lot of trivial problems like those mentioned.

READ MORE BEFORE IMPLEMENTING

Therefore, before building your own hydroponic system, it is always a good idea to begin with reading hydroponics beginner's guides. Take your time to read and fully understand the concept as well as the benefits and drawbacks of different types of hydroponic systems. Personally, I recommend reading at least three books because different books focus on different aspects of hydroponics based on authors' experiences. Besides reading guides, you can also get valuable information by visiting discussion forums related to hydroponic gardening on the internet. Once you have enough information and most importantly understand you own needs or requirements, then you can start developing your own hydroponic gardening system and have fun with it.

PROBLEMS

Methodology cal interventions generate tremendous concern and anxiety in the field of agriculture, though the same interventions in human medicine are accepted without much ado. Apparently the difference in perception is caused by the risks as well as uncertainties involved in introducing Genetically Modified Organisms (GMOs) in the nature. Part of the problem is also that the available information about various risks has not been shared as widely as was necessary to influence public opinion. Finally, the increasing corporate control over biotechnological alternatives also generates tremendous anxiety in the mind of consumers.

In countries such as India where a very large research programme on transgenics is underway in public sector, the popular perceptions are obviously quite different than in Europe where such is perhaps not the case. It is not my argument that public sector cannot be irresponsible or mask information.

However, the individual motivations and the permeability of the organizational boundaries could certainly be different in public systems. Another issue which has remained less explored is the relationship between the quantum of risks involved in a technology and its potential pay off in solving urgent food problems. Notwithstanding the fact that large number of transgenic crops have been developed to increase chemical consumption (through herbicide tolerance), there are equally large biotechnological interventions possible which can reduce chemical consumption.

Further, given the adverse environmental consequences of chemical pesticides and insufficiently developed markets in many developing countries, the uncoupled technologies (seed, fertilizer and chemical separately) always impose higher transaction costs compared to the coupled technology (seed with resistance to pests, incorporation of higher nutrients, tolerance to salts or other stresses). In this paper, I tried to develop one more argument for considering biotechnological interventions more parsimoniously. And this is about the attitude towards risk that small farmers take every day in order to survive and mange their resources. If we can match their capabilities to deal with risks and manage uncertainties in the context of biotechnological interventions, we would have made a significant progress in steering the research on biotechnological alternatives in more benign and sustainable directions.

LOGIC OF SURVIVAL AND PERCEPTION OF RISK

The precautionary principle has always been a part of survival algorithms of small farmers in developing societies particularly in risk prone environments such as drought, flood prone areas, mountainous and forest regions. However, the farmers always manage risk aversion in certain markets by taking extreme risks in other resource markets as a part of their portfolio strategies depending upon the access, assurances, and abilities commanded by them. The issue therefore is to understand how households survive by taking risks in a manner that they not only cope with the consequences but also improve their capacity to deal with uncertainties in future.

Thus, risk taking as an input into capacity building for dealing with bigger risks or uncertainties requires a different way of thinking compared to the choice of risk at a level of survival threshold. To define a survival threshold, one can take the analogy of plimsoll line. If a ship tilts beyond the boundary of this line, it may sink. But within the range of this line, it can cope with turbulence. Survival threshold is the limit within which risks are taken. Occasionally farmers gamble, just as countries and corporations do. What we have to see is whether the gamble is worth it, what are the possible consequences for human health and life, dignity and ultimately for the ecosystem health.

LEARNING FROM RISKS

In the mid sixties Indian planners decided to gamble by importing bulk quantities of seeds from Mexico to herald green revolution in India. There were lot of risks, of diseases, weeds and pests. Indian scientists went to Mexico and inspected the fields from which the seeds were collected and the entire operation quickly transformed within two years the total food production from 67 million tonnes to 110 million tonnes.

A country termed as a 'basket case' moved towards food self-reliance. Today it produces more than 200 million tonnes. Income disparities increased initially, sustainability decreased eventually, productivity of inputs declined and the soil and water quality deteriorated due to excessive chemical use. The agrobiodiversity declined drastically during the same period and entirely due to public sector driven dissemination of high yielding varieties. There was hardly any private sector intervention in grain seed sector.

Can we argue therefore that biotechnological risks have to be analysed through the same the same principle, *i.e.*, a reasonable trade off between the known risks and the social pay off through sufficient institutional capability to deal with the consequences? We must of course, learn the right lessons from the green revolution experience. Without appropriate tuning of institutions (credit, irrigation, insurance, prices, etc.) the technology could not have shown the results. The technology I have argued is like words and the institutions symbolize grammar. The biotechnological interventions will need to be supported by simultaneous development of institutions to take care of risks, unforeseen consequences and at the same time ensuring proper precautions.

MAKING TRADE OFF RATIONALLY

How do we ensure that the trade off between known negative externalities caused by use of chemical pesticides and other inputs vis-à-vis some of the unknown externalities likely to be caused by use of bio pesticides or transgenic crops? Is it necessarily ethical to avoid taking risks and subject societies to suffer deprivation merely because of some risks, which are not completely quantifiable? Should we reduce the risks by getting location specific testing done in each country under rigorous conditions and with all the risks fully disclosed? Each country should have the choice to decide whether the risk is worth taking or not.

Once the level of risk is mutually agreed upon after prior informed consent, the responsibility of the global community is to ensure that a proper support system is available to safeguard the interests of technologically backward countries if such a need arises.

The precautionary principle is a valid means of generating responsibility in taking risks. It is not a means to prevent recognition and calibration of the risk. Once the risks are calibrated, it will depend upon the specific socio-economic conditions and cultural milieu, which will determine how much risk, is acceptable at what stage of economic development and with what consequences. Currently, concern with unknown risks is not matched with responsibility for known consequences of chemical pesticides and other environmental risks such as excessive extraction of ground water, decline in biodiversity, etc.

There are several issues in this debate which have remained obscure. For instance, how to link ethical issues in poverty alleviation with ethical concerns in using or not using risky technologies with suspected environmental impacts; similarly how to deal with the risks that are known but are not attended to adequately (adverse envt and health effects of chemical pesticide, there are hardly any hoardings in most developing countries telling farm workers how to use chemical pesticides safely or deal with the effects caused already) vi-a-vis the ones which are likely to cause the adverse effect in future; what is ethical basis of differential norms of disclosure by the same corporation in developed countries vis-à-vis developed countries; how do we deal with ethical basis of not allocating sufficient research resources to tackle the problems of low productivity in rainfed regions; how to deal with anxieties and fears generated by the larger corporate control of biotechnological research which has not been the case in conventional research; what are the peculiarities of processing complex information in dealing with biotechnological risks compared to other kinds of risks.

An approach to evaluate the risks on ethical, economic, equity and environmental grounds taking into account the prior experience in dealing with different kinds of technologies in a given society is needed. The question is whether the precautionary principle is better used as a tool with which to stop uncharted action, or as a motivation by which to chart those actions contemplated or taken. My preference is for the latter alternative.

2

Nano Processes

RESEARCH

Until the sixteenth century, human inquiry was primarily based on introspection. The way to know things was to turn inward and use logic to seek the truth. This paradigm had endured for a millennium and was a well-established conceptual framework for understanding the world. The seeker of knowledge was an integral part of the inquiry process.

A profound change occurred during the sixteenth and seventeenth centuries. Copernicus, Kepler, Galileo, Descartes, Bacon, Newton, and Locke presented new ways of examining nature. Our method of understanding the world came to rely on measurement and quantification. Mathematics replaced introspection as the key to supreme truths. The *Scientific Revolution* was born.

Objectivity became a critical component of the new scientific method. The investigator was an observer, rather than a participant in the inquiry process. A mechanistic view of the universe evolved. We believed that we could understand the whole by performing an examination of the individual parts. Experimentation and deduction became the tools of the scholar. For two hundred years, the new paradigm slowly evolved to become part of the reality framework of society. The *Age of Enlightenment* had arrived.

Scientific research methodology was very successful at explaining natural phenomena. It provided a systematic way of knowing. Western philosophers embraced this new structure of inquiry. Eastern philosophy continued to stress the importance of the one seeking knowledge. By the beginning of the twentieth century, a complete schism had occurred. Western and Eastern philosophies were mutually exclusive and incompatible.

Then something remarkable happened. Einstein's proposed that the observer was not separate from the phenomena being studied. Indeed, his theory of relativity actually stressed the role of the observer. Quantum mechanics carried this a step further and stated that the act of observation could change the thing being observed. The researcher was not simply an observer, but in fact, was an integral part of the process. In physics, Western and Eastern philosophies have met.

This idea has not been incorporated into the standard social science research model, and today's social science community see themselves as objective observers of the phenomena being studied. However, "it is an established principle of measurement that instruments react with the things they measure." The concept of *instrument reactivity* states that an instrument itself can disturb the thing being measured.

PROBLEMS

All research begins with a question. Intellectual curiosity is often the foundation for scholarly inquiry. Some questions are not testable. The classic philosophical example is to ask, "How many angels can dance on the head of a pin?" While the question might elicit profound and thoughtful revelations, it clearly cannot be tested with an empirical experiment. Prior to Descartes, this is precisely the kind of question that would engage the minds of learned men. Their answers came from within. The modern scientific method precludes asking questions that cannot be empirically tested. If the angels cannot be observed or detected, the question is considered inappropriate for scholarly research.

A paradigm is maintained as much by the process of formulating questions as it is by the answers to those questions. By excluding certain types of questions, we limit the scope of our thinking. It is interesting to note, however, that modern physicists have began to ask the same kinds of questions posed by the Eastern philosophers. "Does a tree falling in the forest make a sound if nobody is there to hear it?" This seemingly trivial question is at the heart of the observer/observed dichotomy. In fact, quantum mechanics predicts that this kind of question cannot be answered with complete certainty. It is the beginning of a new paradigm.

Defining the goals and objectives of a research project is one of the most important steps in the research process. Clearly stated goals keep a research project focused. The process of goal definition usually begins by writing down the broad and general goals of the study. As the process continues, the goals become more clearly defined and the research issues are narrowed. Exploratory research (*e.g.*, literature reviews, talking to people, and focus groups) goes hand-in-hand with the goal clarification process. The literature review is especially important because it obviates the need to *reinvent the wheel* for every new research question. More importantly, it gives researchers the opportunity to build on each others work. The research question itself can be stated as a hypothesis. A hypothesis is simply the investigator's belief about a problem. Typically, a researcher formulates an opinion during the literature review process. The process of reviewing other scholar's work often clarifies the theoretical issues associated with the research question. It also can help to elucidate the significance of the issues to the research community. The hypothesis is converted into a null hypothesis in order to make it testable. "The only way to test a hypothesis is to eliminate alternatives of the hypothesis." (Anderson, 1966, p.9) Statistical techniques will enable us to reject a null hypothesis, but they do not provide us with a way to accept a hypothesis. Therefore, all hypothesis testing is indirect.

CREATING THE RESEARCH DESIGN

Defining a research problem provides a format for further investigation. A well-defined problem points to a method of investigation. There is no one best method of research for all situations. Rather, there are a wide variety of techniques for the researcher to choose from. Often, the selection of a technique involves a series of trade-offs. For example, there is often a trade-off between cost and the quality of information obtained. Time constraints sometimes force a trade-off with the overall research design. Budget and time constraints must always be considered as part of the design process (Walonick, 1993). Many authors have categorized research design as either *descriptive* or *causal*. Descriptive studies are meant to answer the questions of who, what, where, when and how. Causal studies are undertaken to determine how one variable affects another. McDaniel and Gates (1991) state that the two characteristics that define causality are *temporal sequence* and *concomitant variation*.

The word *causal* may be a misnomer. The mere existence of a temporal relationship between two variables does not prove or even imply that A causes B. It is never possible to provecausality. At best, we can theorize about causality based on the relationship between two or more variables, however, this is prone to misinterpretation. Personal bias can lead to totally erroneous statements. For example, Blacks often score lower on I.Q. scores than their White counterparts. It would be irresponsible to conclude that ethnicity causes high or low I.Q. scores. In social science research, making false assumptions about causality can delude the researcher into ignoring other (more important) variables.

There are three basic methods of research:

- Survey,
- Observation, and
- Experiment (McDaniel and Gates, 1991).

Each method has its advantages and disadvantages. The *survey* is the most common method of gathering information in the social sciences. It can be a face-to-face interview, telephone, or mail survey. A personal interview is one of the best methods obtaining personal, detailed, or in-depth information. It usually involves a lengthy questionnaire that the interviewer fills out while asking questions. It allows for extensive probing by the interviewer and gives respondents the ability to elaborate their answers. Telephone interviews are similar to face-to-face interviews. They are more efficient in terms of time and cost, however, they are limited in the amount of in-depth probing that can be accomplished, and the amount of time that can be allocated to the interview. A mail survey is generally the most cost effective interview method. The researcher can obtain opinions, but trying to meaningfully probe opinions is very difficult.

Observation research monitors respondents' actions without directly interacting with them. It has been used for many years by A.C. Nielsen to monitor television viewing habits. Psychologists often use one-way mirrors to study behaviour. Social scientists often study societal and group behaviours by simply observing them. The fastest growing form of observation research has been made possible by the bar code scanners at cash registers, where purchasing habits of consumers can now be automatically monitored and summarized.

In an *experiment*, the investigator changes one or more variables over the course of the research. When all other variables are held constant (except the one being manipulated), changes in the dependent variable can be explained by the change in the independent variable. It is usually very difficult to control all the variables in the environment. Therefore, experiments are generally restricted to laboratory models where the investigator has more control over all the variables.

SAMPLING

It is incumbent on the researcher to clearly define the target population. There are no strict rules to follow, and the researcher must rely on logic and judgment. The population is defined in keeping with the objectives of the study.

Sometimes, the entire population will be sufficiently small, and the researcher can include the entire population in the study. This type of research is called a census study because data is gathered on every member of the population. Usually, the population is too large for the researcher to attempt to survey all of its members. A small, but carefully chosen sample can be used to represent the population. The sample reflects the characteristics of the population from which it is drawn.

Sampling methods are classified as either probability or nonprobability. In probability samples, each member of the population has a known probability of being selected. Probability methods include random sampling, systematic sampling, and stratified sampling. In nonprobability sampling, members are selected from the population in some nonrandom manner. These include convenience sampling, judgment sampling, quota sampling, and snowball sampling. The other common form of nonprobability sampling occurs by accident when the researcher inadvertently introduces nonrandomness into the sample selection process. The advantage of probability sampling is that sampling error can be calculated. Sampling error is the degree to which a sample might differ from the population. When inferring to the population, results are reported plus or minus the sampling error. In nonprobability sampling, the degree to which the sample differs from the population remains unknown. (McDaniel and Gates, 1991)

Random sampling is the purest form of probability sampling. Each member of the population has an equal chance of being selected. When there are very large populations, it is often difficult or impossible to identify every member of the population, so the pool of available subjects becomes biased. Random sampling is frequently used to select a specified number of records from a computer file.

Systematic sampling is often used instead of random sampling. It is also called an Nth name selection technique. After the required sample size has been calculated, every Nth record is selected from a list of population members. As long as the list does not contain any hidden order, this sampling method is as good as the random sampling method. Its only advantage over the random sampling technique is simplicity.

Stratified sampling is commonly used probability method that is superior to random sampling because it reduces sampling error.

A stratum is a subset of the population that share at least one common characteristic. The researcher first identifies the relevant stratums and their actual representation in the population. Random sampling is then used to select subjects for each stratum until the number of subjects in that stratum is proportional to its frequency in the population.

Convenience sampling is used in exploratory research where the researcher is interested in getting an inexpensive approximation of the truth. As the name implies, the sample is selected because they are convenient. This nonprobability method is often used during preliminary research efforts to get a gross estimate of the results, without incurring the cost or time required to select a random sample.

Judgment sampling is a common nonprobability method. The researcher selects the sample based on judgment. This is usually and extension of convenience sampling. For example, a researcher may decide to draw the entire sample from one "representative" city, even though the population includes all cities. When using this method, the researcher must be confident that the chosen sample is truly representative of the entire population.

Quota sampling is the nonprobability equivalent of stratified sampling. Like stratified sampling, the researcher first identifies the stratums and their proportions as they are represented in the population. Then convenience or judgment sampling is used to select the required number of subjects from each stratum. This differs from stratified sampling, where the stratums are filled by random sampling.

Snowball sampling is a special nonprobability method used when the desired sample characteristic is rare. It may be extremely difficult or cost prohibitive to locate respondents in these situations. Snowball sampling relies on referrals from initial subjects to generate additional subjects. While this technique can dramatically lower search costs, it comes at the expense of introducing bias because the technique itself reduces the likelihood that the sample will represent a good cross section from the population.

DATA COLLECTION

There are very few hard and fast rules to define the task of data collection. Each research project uses a data collection technique appropriate to the particular research methodology. The two primary goals for both quantitative and qualitative studies are to maximize response and maximize accuracy.

When using an outside data collection service, researchers often *validate* the data collection process by contacting a percentage of the respondents to verify that they were actually interviewed. Data *editing* and *cleaning* involves the process of checking for inadvertent errors in the data. This usually entails using a computer to check for out-of-bounds data.

Quantitative studies employ deductive logic, where the researcher starts with a hypothesis, and then collects data to confirm or refute the hypothesis. *Qualitative* studies use inductive logic, where the researcher first designs a study and then develops a hypothesis or theory to explain the results of the analysis.

Quantitative analysis is generally fast and inexpensive. A wide assortment of statistical techniques are available to the researcher. Computer software is readily available to provide both basic and advanced multivariate analysis. The researcher simply follows the preplanned analysis process, without making subjective decisions about the data. For this reason, quantitative studies are usually easier to execute than qualitative studies.

Qualitative studies nearly always involve in-person interviews, and are therefore very labour intensive and costly. They rely heavily on a researcher's ability to exclude personal biases. The interpretation of qualitative data is often highly subjective, and different researchers can reach different conclusions from the same data. However, the goal of qualitative research is to develop a hypothesis—not to test one. Qualitative studies have merit in that they provide broad, general theories that can be examined in future research.

DATA ANALYSIS

Modern computer software has made the analysis of quantitative data a very easy task. It is no longer incumbent on the researcher to know the formulas needed to calculate the desired statistics. However, this does not obviate the need for the researcher to understand the theoretical and conceptual foundations of the statistical techniques. Each statistical technique has its own assumptions and limitations. Considering the ease in which computers can calculate complex statistical problems, the danger is that the researcher might be unaware of the assumptions and limitations in the use and interpretation of a statistic.

REPORTING THE RESULTS

The most important consideration in preparing any research report is the nature of the audience. The purpose is to communicate information, and therefore, the report should be prepared specifically for the readers of the report. Sometimes the format for the report will be defined for the researcher (*e.g.*, a dissertation), while other times, the researcher will have complete latitude regarding the structure of the report. At a minimum, the report should contain an abstract, problem statement, methods section, results section, discussion of the results, and a list of references.

VALIDITY AND RELIABILITY

- *Validity* refers to the accuracy or truthfulness of a measurement. Are we measuring what we think we are? "Validity itself is a simple concept, but the determination of the validity of a measure is elusive".
- *Face validity* is based solely on the judgment of the researcher. Each question is scrutinized and modified until the researcher is satisfied that it is an accurate measure of the desired construct. The determination of face validity is based on the subjective opinion of the researcher.
- *Content validity* is similar to face validity in that it relies on the judgment of the researcher. However.

where face validity only evaluates the individual items on an instrument, content validity goes further in that it attempts to determine if an instrument provides adequate coverage of a topic. Expert opinions, literature searches, and pretest open-ended questions help to establish content validity.

- *Criterion-related validity* can be either predictive or concurrent. When a dependent/independent relationship has been established between two or more variables, criterion-related validity can be assessed. A mathematical model is developed to be able to predict the dependent variable from the independent variable(s). *Predictive validity* refers to the ability of an independent variable (or group of variables) to predict a future value of the dependent variable. *Concurrent validity* is concerned with the relationship between two or more variables at the same point in time.
- *Construct validity* refers to the theoretical foundations underlying a particular scale or measurement. It looks at the underlying theories or constructs that explain a phenomena. This is also quite subjective and depends heavily on the understanding, opinions, and biases of the researcher.
- *Reliability* is synonymous with repeatability. A measurement that yields consistent results over time is said to be reliable. When a measurement is prone to random error, it lacks reliability. The reliability of an instrument places an upper limit on its validity. A measurement that lacks reliability will necessarily be invalid. There are three basic methods to test reliability: test-retest, equivalent form, and internal consistency.
- A *test-retest* measure of reliability can be obtained by administering the same instrument to the same group of people at two different points in time. The degree to which both administrations are in agreement is a measure of the reliability of the instrument. This

technique for assessing reliability suffers two possible drawbacks. First, a person may have changed between the first and second measurement. Second, the initial administration of an instrument might in itself induce a person to answer differently on the second administration.

The second method of determining reliability is called the *equivalent-form* technique. The researcher creates two different instruments designed to measure identical constructs. The degree of correlation between the instruments is a measure of equivalent-form reliability. The difficulty in using this method is that it may be very difficult (and/or prohibitively expensive) to create a totally equivalent instrument.

The most popular methods of estimating reliability use measures of *internal consistency*. When an instrument includes a series of questions designed to examine the same construct, the questions can be arbitrarily split into two groups. The correlation between the two subsets of questions is called the *split-half* reliability.

The problem is that this measure of reliability changes depending on how the questions are split. A better statistic, known as Chronbach's alpha, is based on the mean (absolute value) interitem correlation for all possible variable pairs. It provides a conservative estimate of reliability, and generally represents "the lower bound to the reliability of an unweighted scale of items". For dichotomous nominal data, the KR-20 is used instead of Chronbach's alpha.

VARIABILITY AND ERROR

Most research is an attempt to understand and explain *variability*. When a measurement lacks variability, no statistical tests can be (or need be) performed. Variability refers to the dispersion of scores.

Ideally, when a researcher finds differences between respondents, they are due to true difference on the variable being measured. However, the combination of systematic and random errors can dilute the accuracy of a measurement.

Systematic error is introduced through a constant bias in a measurement. It can usually be traced to a fault in the sampling procedure or in the design of a questionnaire. *Random error* does not occur in any consistent pattern, and it is not controllable by the researcher.

Scientific research involves the formulation and testing of one or more hypotheses. A hypothesis cannot be proved directly, so a null hypothesis is established to give the researcher an indirect method of testing a theory. Sampling is necessary when the population is too large, or when the researcher is unable to investigate all members of the target group.

Random and systematic sampling are the best methods because they guarantee that each member of the population will have an known non-zero chance of being selected. The mathematical reliability (repeatability) of a measurement, or group of measurements, can be calculated, however, validity can only be implied by the data, and it is not directly verifiable. Social science research is generally an attempt to explain or understand the variability in a group of people.

STEPS

STEPS IN THE RESEARCH DESIGN PROCESS

The steps in the design process interact and often occur simultaneously. For example, the design of a measurement instrument is influenced by the type of analysis that will be conducted. However, the type of analysis is also influenced by the specific characteristics of themeasurement instrument.

Define the Research Problem

Problem definition is the most critical part of the research process. Research problem definition involves specifying the information needed by management. Unless the problem is properly defined, the information produced by the research process is unlikely to have any value. Coca-Cola Company researchers utilized a very sound research design to collect information on taste preferences. Unfortunately for

Coca-Cola, taste preferences are only part of what drives the soft drink purchase decision.

Research problem definition involves four interrelated steps:

1. Management problem/opportunity clarification,
2. Situation analysis,
3. Model development, and
4. Specification of information requirements.

The basis goal of problem clarification is to ensure that the decision maker's initial description of the management decision is accurate and reflects the appropriate area of concern for research. If the wrong management problem is translated into a research problem, the probability of providing management with useful information is low.

Situation Analysis

The situation analysis focuses on the variables that have produced the stated managementproblem or opportunity. The factors that have led to the problem/opportunity manifestations and the factors that have led to management's concern should be isolated.

A situation analysis of the retail trade outflow problem revealed, among other things, that:

- The local population had grown 25 per cent over the previous five years,
- Buying power per capita appeared to be growing at the national rate of 3 per cent a year, and
- Local retail sales of nongrocery items had increased approximately 20 per cent over the past five years.

Thus, the local retailers sales are clearly not keeping pace with the potential in the area.

Estimate the Value of the Information

A decision maker normally approaches a problem with some information. If the problem is, say, whether a new product should be introduced, enough information will normally have been accumulated through past experience with other decisions concerning the introduction of new products and from various other sources to allow some preliminary judgments to be formed about the desirability of introducing the product in question. There will rarely be sufficient confidence in these judgments that additional information relevant to the decision would not be accepted if it were available without cost or delay. There might be enough confidence, however, that there would be an unwillingness to pay very much or wait very long for the added information.

Select the Data Collection Approach

There are three basic data collection approaches in marketing research:

1. Secondary data,
2. Survey data, and
3. Experimental data.

Secondary data were collected for some purpose other than helping to solve the current problem, whereas primary data are collected expressly to help solve the problem at hand.

Select the Measurement Technique

There are four basic measurement techniques used in marketing research:

- Questionnaires,
- Attitude scales,
- Observation, and
- Depth interviews and projective techniques.

Primary Measurement Techniques

I. *Questionnaire*: A formalized instrument for asking information directly from a respondent concerning behaviour, demographic characteristics, level of knowledge, and/or attitudes, beliefs, and feelings.

II. *Attitude Scales:* a formalized instrument for eliciting self-reports of beliefs and feelings concerning an object(s).

   A. *Rating Scales:* Require the respondent to place the object being rated at some point along a numerically valued continuum or in one of a numerically ordered series of categories.

B. *Composite Scales:* Require the respondents to express a degree of belief concerning various attributes of the object such that the attitude can be inferred from the pattern of responses.

C. *Perceptual maps:* derive the components or characteristics an individual uses in comparing similar objects and provide a score for each object on each characteristic.

D. *Conjoint analysis*: Derive the value an individual assigns to various attributes of a product.

I. *Observation:* The direct examination of behaviour, the results of behaviour, or physiological changes.

II. *Projective Techniques and Depth Interview:* Designed to gather information that respondents are either unable or unwilling to provide in response to direct questioning.

A. *Projective Techniques*: Allow respondents to project or express their own feelings as a characteristic of someone or something else.

B. *Depth Interviews:* Allow individuals to express themselves without any fear of disapproval, dispute, or advice from the interviewer.

Select the Sample

Most marketing studies involve a sample or subgroup of the total population relevant to the problem, rather than a census of the entire group.

Select the Model of Analysis

It is imperative that the researcher select the analytic techniques prior to collecting the data. Once the analytic techniques are selected, the researcher should generate fictional responses (dummy data) to the measurement instrument. These dummy data are then analysed by the analytic techniques selected to ensure that the results of this analysis will provide the information required by the problem at hand.

Evaluate the Ethics of the Research

It is essential that marketing researchers restrict their

research activities to practices that are ethically sound. Ethically sound research considers the interests of the general public, the respondents, the client and the research profession as well as those of the researcher.

Estimate Time and Financial Requirements

The program evaluation review technique (PERT) coupled with the critical path method (CPM) offers a useful aid for estimating the resources needed for a project and clarifying the planning and control process. PERT involves dividing the total research project into its smallest component activities, determining the sequence in which these activities must be performed, and attaching a time estimate for each activity. These activities and time estimates are presented in the form of a flow chart that allow a visual inspection of the overall process. The time estimates allow one to determine the critical path through the chart – that series of activities whose delay will hold up the completion of the project.

Prepare the Research Proposal

The research design process provides the researcher with a blueprint, or guide, for conducting and controlling the research project. The blueprint is written in the form of a research proposal. A written research proposal should precede any research project.

VARIOUS METHODS OF RESEARCH DESIGN

The research method you will follow, is directly connected to your problem statement and goal of research. Because the research goal and problem may vary different methods of research can be utilized. Research is a purposeful, precise and systematic search for new knowledge, skills, attitudes and values, or for the re-interpretation of existing knowledge, skills, attitudes and values.

The various kinds of human science research can be subdivided according to three criteria:

1. *The measure of generality and applicability*:
   - Basic research

- Applied research
- In-service research
- Action research

2. *The level of ordering*:
   - Descriptive research
   - Prophetic research
   - Diagnostic research
3. *The measure of control by researchers*:
   - Library research
   - Field research
   - Laboratory research
   - *Methods of research:*
     - Exploratory research
     - Experimental research
     - Ex post facto research
     - Correlation research
     - Descriptive research
     - Testing research
     - Case studies
     - Sociometric research
     - Instrumental-nomological research
     - Interpretative-theoretical research

     Other names given to research:
     - Micro-study
     - Macro-study
     - Longitudinal (diachronic) study
     - Cross-sectional (synchronic) study
     - Pilot study

AN OVERVIEW OF SOME RESEARCH METHODS

ACTION RESEARCH

Action research is regarded as research that is normally carried out by practitioners (persons that stand in the field of work). It is a method *par excel lance* for instructors/trainers. It enables the researcher to investigate a specific problem that exists in practice. According to Landman this requires that the researcher should be involved in the actions that take place. A further refinement of this type of research is that the results obtained from the research should be relevant. to the practice. In other words it should be applicable immediately. This means that the, researcher, as expert, and the person standing in the practice, jointly decide on the formulation of research procedures, allowing the problem to be solved.

Action research is characterized according to by the following four features: Problem-aimed research focuses on a special situation in practice. Seen in research context, action research is aimed at a specific problem recognizable in practice, and of which the outcome problem solving) is immediately applicable in practice.

- *Collective participation:* A second characteristic is that all participants (for instance the researchers and persons standing in the practice) form an integral part of action research with the exclusive aim to assist in solving the identified problem.
- *Type of empirical research:* Thirdly, action research is characterized as a means to change the practice while the research is going on.

Outcome of research can not be generalized. Lastly, action research is characterized by the fact that problem solving, seen as renewed corrective actions, can not be generalized, because it should comply with the criteria set for scientific character. An Overview of the Methodological Approach of Action Research

INTRODUCTION

"If you want it done right, you may as well do it yourself." This aphorism may seem appropriate if you are a picky housekeeper, but more and more people are beginning to realise it can also apply to large corporations, community development projects, and even national governments. Such entities exist increasingly in an interdependent world, and are relying on Action Research as a means of coming to grips with their constantly changing and turbulent environments.

This paper will answer the question "What is Action Research?", giving an overview of its processes and principles, stating when it is appropriate to use, and situating it within a praxis research paradigm. The evolution of the approach will be described, including the various kinds of action research being used today. The role of the action researcher will be briefly mentioned, and some ethical considerations discussed. The tools of the action researcher, particularly that of the use of search conferences, will be explained. Finally three case studies will be briefly described, two of which pertain to action research projects involving information technology, a promising area needing further research.

WHAT IS ACTION RESEARCH

Action research is known by many other names, including participatory research, collaborative inquiry, emancipatory research, action learning, and contextual action research, but all are variations on a theme.

Put simply, action research is "learning by doing"-a group of people identify a problem, do something to resolve it, see how successful their efforts were, and if not satisfied, try again. While this is the essence of the approach, there are other key attributes of action research that differentiate it from common problem-solving activities that we all engage in every day.

A more succinct definition is, action research...aims to contribute both to the practical concerns of people in an immediate problematic situation and to further the goals of social science simultaneously. Thus, there is a dual commitment in action research to study a system and concurrently to collaborate with members of the system in changing it in what is together regarded as a desirable direction. Accomplishing this twin goal requires the active collaboration of researcher and client, and thus it stresses the importance of co-learning as a primary aspect of the research process.

What separates this type of research from general professional practices, consulting, or daily problem-solving is the emphasis on scientific study, which is to say the researcher studies the problem systematically and ensures the intervention is informed by theoretical considerations. Much of the researcher's time is spent on refining the methodological tools to suit the exigencies of the situation, and on collecting, analyzing, and presenting data on an ongoing, cyclical basis.

Several attributes separate action research from other types of research. Primary is its focus on turning the people involved into researchers, too-people learn best, and more willingly apply what they have learned, when they do it themselves. It also has a social dimension-the research takes place in real-world situations, and aims to solve real problems. Finally, the initiating researcher, unlike in other disciplines, makes no attempt to remain objective, but openly acknowledges their bias to the other participants.

The Action Research Process

Stephen Kemmis has developed a simple model of the cyclical nature of the typical action research process. Each cycle has four steps: plan, act, observe, reflect.

Gerald Susman gives a somewhat more elaborate listing. He distinguishes five phases to be conducted within each research cycle. Initially, a problem is identified and data is collected for a more detailed diagnosis. This is followed by a collective postulation of several possible solutions, from which a single plan of action emerges and is implemented. Data on the results of the intervention are collected and analysed, and the findings are interpreted in light of how successful the action has been. At this point, the problem is re-assessed and the process begins another cycle.

Principles of Action Research

What gives action research its unique flavour is the set of principles that guide the research.

Winter (1989) provides a comprehensive overview of six key principles:

1. *Reflexive critique*: An account of a situation, such as notes, transcripts or official documents, will make implicit claims to be authoritative, *i.e.,* it implies that

it is factual and true. Truth in a social setting, however, is relative to the teller. The principle of reflective critique ensures people reflect on issues and processes and make explicit the interpretations, biases, assumptions and concerns upon which judgments are made. In this way, practical accounts can give rise to theoretical considerations.

2. *Dialectical critique*: Reality, particularly social reality, is consensually validated, which is to say it is shared through language. Phenomena are conceptualized in dialogue, therefore a dialectical critique is required to understand the set of relationships both between the phenomenon and its context, and between the elements constituting the phenomenon. The key elements to focus attention on are those constituent elements that are unstable, or in opposition to one another. These are the ones that are most likely to create changes.
3. *Collaborative Resource*: Participants in an action research project are co-researchers. The principle of collaborative resource presupposes that each person's ideas are equally significant as potential resources for creating interpretive categories of analysis, negotiated among the participants. It strives to avoid the skewing of credibility stemming from the prior status of an idea-holder. It especially makes possible the insights gleaned from noting the contradictions both between many viewpoints and within a single viewpoint
4. *Risk*: The change process potentially threatens all previously established ways of doing things, thus creating psychic fears among the practitioners. One of the more prominent fears comes from the risk to ego stemming from open discussion of one's interpretations, ideas, and judgments. Initiators of action research will use this principle to allay others' fears and invite participation by pointing out that they, too, will be subject to the same process, and

that whatever the outcome, learning will take place.

5. *Plural Structure*: The nature of the research embodies a multiplicity of views, commentaries and critiques, leading to multiple possible actions and interpretations. This plural structure of inquiry requires a plural text for reporting. This means that there will be many accounts made explicit, with commentaries on their contradictions, and a range of options for action presented. A report, therefore, acts as a support for ongoing discussion among collaborators, rather than a final conclusion of fact.
6. *Theory, Practice, Transformation*: For action researchers, theory informs practice, practice refines theory, in a continuous transformation. In any setting, people's actions are based on implicitly held assumptions, theories and hypotheses, and with every observed result, theoretical knowledge is enhanced. The two are intertwined aspects of a single change process. It is up to the researchers to make explicit the theoretical justifications for the actions, and to question the bases of those justifications. The ensuing practical applications that follow are subjected to further analysis, in a transformative cycle that continuously alternates emphasis between theory and practice.

When is Action Research used

Action research is used in real situations, rather than in contrived, experimental studies, since its primary focus is on solving real problems. It can, however, be used by social scientists for preliminary or pilot research, especially when the situation is too ambiguous to frame a precise research question. Mostly, though, in accordance with its principles, it is chosen when circumstances require flexibility, the involvement of the people in the research, or change must take place quickly or holistically.

It is often the case that those who apply this approach are practitioners who wish to improve understanding of their practice, social change activists trying to mount an action campaign, or, more likely, academics who have been invited into an organization (or other domain) by decision-makers aware of a problem requiring action research, but lacking the requisite methodological knowledge to deal with it.

Situating Action Research in a Research Paradigm

Positivist Paradigm

The main research paradigm for the past several centuries has been that of Logical Positivism. This paradigm is based on a number of principles, including: a belief in an objective reality, knowledge of which is only gained from sense data that can be directly experienced and verified between independent observers. Phenomena are subject to natural laws that humans discover in a logical manner through empirical testing, using inductive and deductive hypotheses derived from a body of scientific theory. Its methods rely heavily on quantitative measures, with relationships among variables commonly shown by mathematical means. Positivism, used in scientific and applied research, has been considered by many to be the antithesis of the principles of action research.

Interpretive Paradigm

Over the last half century, a new research paradigm has emerged in the social sciences to break out of the constraints imposed by positivism. With its emphasis on the relationship between socially-engendered concept formation and language, it can be referred to as the Interpretive paradigm. Containing such qualitative methodological approaches as phenomenology, ethnography, and hermeneutics, it is characterized by a belief in a socially constructed, subjectively-based reality, one that is influenced by culture and history. Nonetheless it still retains the ideals of researcher objectivity, and researcher as passive collector and expert interpreter of data.

Paradigm of Praxis

Though sharing a number of perspectives with the

interpretive paradigm, and making considerable use of its related qualitative methodologies, there are some researchers who feel that neither it nor the positivist paradigms are sufficient epistemological structures under which to place action research. Rather, a paradigm of Praxis is seen as where the main affinities lie.

Praxis, a term used by Aristotle, is the art of acting upon the conditions one faces in order to change them. It deals with the disciplines and activities predominant in the ethical and political lives of people. Aristotle contrasted this with Theoria-those sciences and activities that are concerned with knowing for its own sake.

Both are equally needed he thought. That knowledge is derived from practice, and practice informed by knowledge, in an ongoing process, is a cornerstone of action research. Action researchers also reject the notion of researcher neutrality, understanding that the most active researcher is often one who has most at stake in resolving a problematic situation.

EVOLUTION OF ACTION RESEARCH

Origins in late 1940s

Kurt Lewin is generally considered the 'father' of action research. A German social and experimental psychologist, and one of the founders of the Gestalt school, he was concerned with social problems, and focused on participative group processes for addressing conflict, crises, and change, generally within organizations. Initially, he was associated with the Center for Group Dynamics at MIT in Boston, but soon went on to establish his own National Training Laboratories.

Lewin first coined the term 'action research' in his 1946 paper "Action Research and Minority Problems" characterizing Action Research as "a comparative research on the conditions and effects of various forms of social action and research leading to social action", using a process of "a spiral of steps, each of which is composed of a circle of planning, action, and fact-finding about the result of the action".

Eric Trist, another major contributor to the field from that immediate post-war era, was a social psychiatrist whose group at the Tavistock Institute of Human Relations in London engaged in applied social research, initially for the civil repatriation of German prisoners of war. He and his colleagues tended to focus more on large-scale, multi-organizational problems.

Both Lewin and Trist applied their research to systemic change in and between organizations. They emphasized direct professional-client collaboration and affirmed the role of group relations as basis for problem-solving. Both were avid proponents of the principle that decisions are best implemented by those who help make them.

Current Types of Action Research

By the mid-1970s, the field had evolved, revealing 4 main 'streams' that had emerged: traditional, contextural (action learning), radical, and educational action research.

Traditional Action Research

Traditional Action Research stemmed from Lewin's work within organizations and encompasses the concepts and practices of Field Theory, Group Dynamics, T-Groups, and the Clinical Model. The growing importance of labour-management relations led to the application of action research in the areas of Organization Development, Quality of Working Life (QWL), Socio-technical systems (*e.g.*, Information Systems), and Organizational Democracy. This traditional approach tends towards the conservative, generally maintaining the status quo with regards to organizational power structures.

Contextural Action Research (Action Learning)

Contextural Action Research, also sometimes referred to as Action Learning, is an approach derived from Trist's work on relations between organizations. It is contextural, insofar as it entails reconstituting the structural relations among actors in a social environment; domain-based, in that it tries to involve all affected parties and stakeholders; holographic, as each participant understands the working of the whole; and it stresses that participants act as project designers and co-researchers. The concept of organizational ecology, and the use of search conferences come out of contextual action research, which is more of a liberal philosophy, with social transformation occurring by consensus and normative incrementalism.

Radical Action Research

The Radical stream, which has its roots in Marxian 'dialectical materialism' and the praxis orientations of Antonio Gramsci, has a strong focus on emancipation and the overcoming of power imbalances. Participatory Action Research, often found in liberationist movements and international development circles, and Feminist Action Research both strive for social transformation via an advocacy process to strengthen peripheral groups in society.

Educational Action Research

A fourth stream, that of Educational Action Research, has its foundations in the writings of John Dewey, the great American educational philosopher of the 1920s and 30s, who believed that professional educators should become involved in community problem-solving. Its practitioners, not surprisingly, operate mainly out of educational institutions, and focus on development of curriculum, professional development, and applying learning in a social context. It is often the case that university-based action researchers work with primary and secondary school teachers and students on community projects.

ACTION RESEARCH TOOLS

Action Research is more of a holistic approach to problem-solving, rather than a single method for collecting and analyzing data. Thus, it allows for several different research tools to be used as the project is conducted. These various methods, which are generally common to the qualitative research paradigm, include: keeping a research journal, document collection and analysis, participant observation recordings, questionnaire surveys, structured and unstructured interviews, and case studies.

The Search Conference

Of all of the tools utilized by action researchers, the one that has been developed exclusively to suit the needs of the action research approach is that of the search conference, initially developed by Eric Trist and Fred Emery at the Tavistock Institute in 1959, and first implemented for the merger of Bristol-Siddley Aircraft Engines in 1960.

The search conference format has seen widespread development since that time, with variations on Trist and Emery's theme becoming known under other names due to their promotion by individual academics and consultants. These includeDannemiller-Tyson's Interactive Strategic Planning, Marvin Weisbord's Future Search Conference, Dick Axelrod's Conference Model Redesign, Harrison Owen's Open Space, and ICA's Strategic Planning (Rouda 1995).

Search conferences also have been conducted for many different circumstances and participants, including: decision-makers from several countries visioning the "Future of Participative Democracy in the Americas"; practitioners and policymakers in the field of health promotion in Ontario taking charge in an era of cutbacks; and Xerox employees sorting out enterprise re-organization.

Eric Trist Sums up the Process Quite Nicely

Searching...is carried out in groups which are composed of the relevant stakeholders. The group meets under social island conditions for 2-3 days, sometimes as long as five. The opening sessions are concerned with elucidating the factors operating in the wider contextual environment-those producing the meta-problems and likely to affect the future. The content is contributed entirely by the members. The staff are facilitators only. Items are listed in the first instance without criticism in the plenary session and displayed on flip charts which surround the room. The material is discussed in greater depth in small groups and the composite picture checked out in plenary. The group next examines its own organizational setting or settings against this wider background and then proceeds to construct a picture of a desirable future. It is surprising how much agreement there often is. Only when all this has been done is consideration given to action steps.

ACTION RESEARCHER

Upon invitation into a domain, the outside researcher's role is to implement the Action Research method in such a manner as to produce a mutually agreeable outcome for all participants, with the process being maintained by them afterwards. To accomplish this, it may necessitate the adoption of many different roles at various stages of the process, including those of

Planner	Leader
Catalyzer	Facilitator
Teacher	Designer
Listener	Observer
Synthesizer	Reporter

The main role, however, is to nurture local leaders to the point where they can take responsibility for the process. This point is reached they understand the methods and are able to carry on when the initiating researcher leaves.

In many Action Research situations, the hired researcher's role is primarily to take the time to facilitate dialogue and foster reflective analysis among the participants, provide them with periodic reports, and write a final report when the researcher's involvement has ended.

ETHICAL CONSIDERATIONS

Because action research is carried out in real-world circumstances, and involves close and open communication among the people involved, the researchers must pay close attention to ethical considerations in the conduct of their work.

Richard Winter (1996) lists a number of principles:

- "Make sure that the relevant persons, committees and authorities have been consulted, and that the principles guiding the work are accepted in advance by all.
- All participants must be allowed to influence the work, and the wishes of those who do not wish to participate must be respected.
- The development of the work must remain visible and open to suggestions from others.
- Permission must be obtained before making observations or examining documents produced for other purposes.
- Descriptions of others' work and points of view must be negotiated with those concerned before being published.
- The researcher must accept responsibility for maintaining confidentiality."

To this might be added several more points:

- Decisions made about the direction of the research and the probable outcomes are collective
- Researchers are explicit about the nature of the research process from the beginning, including all personal biases and interests
- There is equal access to information generated by the process for all participants
- The outside researcher and the initial design team must create a process that maximizes the opportunities for involvement of all participants.

EXAMPLES OF ACTION RESEARCH PROJECTS

To better illustrate how action research can proceed, three case studies are presented. Action research projects are generally situationally unique, but there are elements in the methods that can be used by other researchers in different circumstances. The first case study, an account taken from the writings of one of the researchers involved, involves a research project to stimulate the development of nature tourism services in the Caribbean. It represents a fairly typical example of an action research initiative. The second and third case studies centre around the use of computer communications, and therefore illustrate a departure from the norm in this regard. They are presented following a brief overview of this potentially promising technical innovation.

Case Study 1: Development of nature tourism in the Windward Islands

In 1991, an action research process was initiated to explore how nature tourism could be instituted on each of the four Windward Islands in the Caribbean-St. Lucia, Grenada, Dominica, and St. Vincent. The government took the lead, for environmental conservation, community-based development, and national economic development purposes. Realizing that the consultation process had to involve many stakeholders, including representatives of several government ministries, environmental and heritage groups, community organizations, women's and youth groups, farmers' cooperatives, and private business, an action research approach was seen as appropriate.

Two action researchers from University, with prior experience in the region, were hired to implement the project, with a majority of the funding coming from the Canadian International Development Agency. Multi-stakeholder national advisory councils were formed, and national project coordinators selected as local project liaisons. Their first main task was to organize a search conference on each island.

The search conferences took place, the outcome of which was a set of recommendations and/or action plans for the carrying out of a number of nature tourism-oriented sub-projects at the local community level. At this point, extended advisory groups were formed on several of the islands, and national awareness activities and community sub-projects were implemented in some cases.

To maintain the process, regional project meetings were held, where project coordinators and key advisory members shared experiences, conducted self-evaluations and developed plans for maintaining the process (*e.g.*, fundraising). One of the more valuable tools for building a sense of community was the use of a videocamera to create a documentary video of a local project.

The outcomes varied. In St. Vincent the research project was highly successful, with several viable local developments instituted. Grenada and St. Lucia showed mixed outcomes, and Dominica was the least successful, the process curtailed by the government soon after the search conference took place. The main difference in the outcomes, it was felt, was in the willingness of the key government personnel to "let go" and allow the process to be jointly controlled by all participants. There is always a risk that this kind of research will empower stakeholders, and change existing power relations, the threat of which is too much for some decision-makers, but if given the opportunity, there are many things that a collaborative group of citizens can accomplish that might not be possible otherwise.

Action Research and Information Technology

In the past ten years or so, there has been a marked increase in the number of organizations that are making use of information technology and computer mediated communications. This has led to a number of convergences between information systems and action research. In some cases, it has been a matter of managers of corporate networks employing action research techniques to facilitate large-scale changes to their information systems. In others, it has been a question of community-based action research projects making use of computer communications to broaden participation.

Much of the action research carried out over the past 40 years has been conducted in local settings with the participants meeting face-to-face with "real-time" dialogue. The emergence of the Internet has led to an explosion of *asynchronous*and *aspatial* group communication in the form of e-mail and computer conferences, and recently, v-mail and video conferencing. While there have been numerous attempts to use this new technology in assisting group learning, both within organizations and among groups in the community [this author has been involved with a dozen or more projects of this kind in the nonprofit sector in Canada alone], there is a dearth of published studies on the use of action research methods in such projects Lau and Hayward (1997), in a recent review of the literature, found that most research on group support systems to date has been in short-term, experimental situations using quantitative methods. There are a few examples, though, of longitudinal studies in naturalistic settings using qualitative methods; of those that did use action research, none studied the use and effects of communication systems in groups and organizations.

We can now to turn to the case studies, both of which are situated in an area in need of more research-that of the use of information technology as a potentially powerful adjunct to action research processes.

Case Study 2: Internet-based collaborative work groups in community health

Lau and Hayward used an action research approach in a study of their own to explore the structuration of Internet-based collaborative work groups. Over a two-year period, the researchers participated as facilitators in three action research cycles of problem-solving among approximately 15 instructors and project staff, and 25 health professionals from various regions striving to make a transition to a more community-based health program. The aim was to explore how Internet-based communications would influence their evolution into a virtual collaborative workgroup.

The first phase was taken up with defining expectations, providing the technology and developing the customised workgroup system. Feedback from participants noted that shorter and more spaced training sessions, with instructions more focused on specific projects would have been more helpful. The next phase saw the full deployment of the system, and the main lesson learned was that the steepness of the learning curve was severely underestimated, with frustrations only minimally satisfied by a great deal of technical support provided by telephone. The final cycle saw the stabilization of the system and the emergence of the virtual groups

The researchers found that those who used the system interactively were more likely to establish projects that were collaborative in nature, and that the lack of high quality information on community healthcare online was a drawback. The participants reported learning a great deal from the initiative.

The interpretations of the study suggest that role clarity, relationship building, information sharing, resource support, and experiential learning are important aspects in virtual group development. There was also a sense that more research was needed on how group support systems can help groups interact with their external environment, as well as on how to enhance the process of learning by group members.

Case Study 3: Computer conferencing in a learning community

Comstock and Fox (1995) have written about their experiences in integrating computer conferencing into a learning community for mid-career working adults attending a Graduate Management Program at University in Seattle. The researchers and their students made use of a dial-up computer conferencing system called Caucus to augment learning outside of monthly classroom weekends. Their findings relate to establishing boundaries to interaction, creating a caring community, and building collaborative learning.

Boundary setting was a matter of both defined membership, *i.e.*, access to particular conferences, and actual participation. The architecture of the online environment was equated to that of a house, in which locked rooms allowed for privacy, but hampered interaction. They suggest some software design changes that would provide more cues and flexibility to improve access and usage.

Relationships in a caring community were fostered by caring talk, personal conversations and story telling. Over time, expressions of personal concern for other participants increased, exemplifying a more tightly-knit group. Playful conversations of a personal nature also improved group relations, as did stories of events in individuals' lives. These processes provided the support and induced the trust needed to sustain the more in-depth collaborative learning taking place.

Students were expected to use the system for collaborative learning using three forms of conversation-dialogue, discussion and critical reflection. Dialogues were enjoined as a result of attempts to relate classroom lessons to personal situations at work, with a better understanding provided by multiple opinions. Discussions, distinguished by the goal of making a group decision or taking an action, required a fair degree of moderation, insofar as participants found it difficult to reach closure. The process of reflecting critically on ideas was also difficult-participants rarely took the time to analyse postings, preferring a more immediate, and more superficial, conversational style.

The authors conclude with four recommendations:

1. Be clear about the purpose of the computer conference and expectations for use;
2. Develop incentives for widespread and continuous participation;
3. Pay attention to affects of the software on the way the system is used for learning; and
4. Teach members of the community how to translate face-to-face collaborative processes to the on-line environment.

Commentary on The need for more Research

The characteristics of the new information technologies, especially that of computer conferencing, which allows group communications to take place outside of the bounds of time and space, have the potential to be well suited to action research. Projects that traditionally have been limited to local, real-time interactions, such as in the case of search conferences, now have the possibility of being conducted online, with the promise of larger-sized groups, more reflexivity, greater geographic reach, and for a longer period of sustained interaction.

The current state of the software architecture, though, does not seem to be sufficient to induce the focused collaboration required. Perhaps this will remain the case until cyberspace becomes as elaborate in contextual cues as our current socio-physical environment. Whatever the eventual outcome of online developments, it is certain that action research and information technologies will continue to converge, and we must be prepared to use action research techniques to better understand and utilize this convergence.

Conclusion

This paper has presented an overview of action research as a methodological approach to solving social problems. The principles and procedures of this type of research, and epistemological underpinnings, were described, along with the evolution of the practice.

Details of a search conference and other tools were given, as was an indication of the roles and ethics involved in the research. The case studies gave concrete examples of projects, particularly in the relatively new area of social deployment of information technologies. Further action research is needed to explore the potential for developing computer-mediated communications in a way that will enhance human interactions.

Historical Research

Historical research, as the term implies, is research based on describing the past. This type of research includes for instance investigations like the recording, analysis and interpretation of events in the past with the purpose of discovering generalizations and deductions that can be useful in understanding the past, the present and to a limited extent, can anticipate the future.

Historians should consequently aspire to getting to the original events that took place and therefore the researcher is dependent on the availability of documentary sources.

According to Klopper collected data for historical research should pass the following test before it can be applied for research purposes namely:

- External evidence or criteria that will account for the authenticity of the information should be included;
- Internal evidence or criteria should be included that will explain the meaning of the data.

Although the chronological sequence of events should be precisely acknowledged, researchers should bear in mind the fact that mere compilation of chronological events is not considered research in itself.

An investigation can only be regarded as scientific research when the researcher interprets the events that took place by pointing out their relationship to the problem investigated, and explaining their meaning.

On account of the voluminous data that historians collect, it is extremely important that attention should be given to a specific plan for the obtaining and organizing of information as well as the retrieval thereof, before this type of research is attempted.

Lastly, it should be mentioned that historical research also encompasses research concerning the origin, development and influence of ideas of the past. As examples, aspects like the origin, development and influence of communism, democracy, capitalism etc can be mentioned. Should you like to do this type of research you can consult the recommended literature listed in the bibliography.

Descriptive Research

The term descriptive is self-explanatory and terminology synonymous to this type of research is: describe, write on, depict. The aim of descriptive research is to verify formulated hypotheses that refer to the present situation in order to elucidate it.

Descriptive research is thus a type of research that is primarily concerned with describing the nature or conditions and degree in detail of the present situation. The emphasis is on describe rather than on judge or interpret.

According to Klopper (1990: 64) researchers who use this method for their research usually aim at:

- Demarcating the population (representative of the universum) by means of perceiving accurately research parameters; and
- Recording in the form of a written report of that which has been perceived.

The aim of the latter is, that when the total record has been compiled, revision of the documents can occur so that the perceptions derived at can be thoroughly investigated. Because the total population (universum) during a specific investigation can not be contemplated as a whole, researchers make use of the demarcation of the population or of the selection of a representative test sample. Test sampling therefore forms an integral part of descriptive research.

In descriptive research the following steps should be included:

- *Problem selection and problem formulation*: The research problem being tested should be explicitly formulated in the form of a question.
- *Literature search*: Intensive literature search regarding the formulated problem enables the researcher to divide the problem into smaller units.
- Problem reduction.
- Hypothesis formulation.
- *Test sampling*: The researcher should determine the size of the test sample.
- *Information retrieval*: The application of appropriate information retrieval techniques to comply with the criteria set for authenticity and competency, is relevant.
- *General planning*: Any research requires sound planning.
- *Report writing*: The report entails the reproduction of factual information, the interpretation of data, conclusions derived from the research and recommendations.

You should make sure that you understand the meaning of the terminology used. Consult the recommended sources for detailed explanations. However, further reference must be made to aspects related to test sampling.

Test Sampling

As mentioned previously, when descriptive research is exposed, demarcation of the population become unavoidable. Test sampling therefor forms an integra! part of this type of research.

Two important questions arise frequently when test sampling is anticipated by researchers, namely:

- How big should the test sample be?
- What is the probability of mistakes occurring in the use of test sampling (instead of the whole population)?

Special care should be taken with the selection of test samples. The results obtained from a survey can never be more authentic than the standard of the population or the representatives of the test sample, according to Klopper. The size of the test sample can also be specified by means of statistics. It is important for the researcher to bear in mind that it is desirable that test sampling be made as large as possible.

The most important criterium that serves as a guideline here, is the extent to which the test sample corresponds with the qualities and characteristics of the general population being investigated. The next three factors should be taken into consideration before a decision is made with regard to the size of the test sample:

- What is the grade of accuracy expected between the test sample and the general population?
- What is the variability of the population? (This, in general terms, is expressed as the standard deviation.)
- What methods should be used in test sampling?

Bias Saying

When you attempt descriptive research, you should take care that the test sample reflects the actual population it represents. The following example holds validity for the latter:

you cannot make a statement regarding all first-year students if you do not include all first-year students in your research. If you do make such a statement, you have to select enrolled first-year students at all the tertiary institutions or a balanced proportional manner, and include the latter when you select your test sample for your research.

Landman points out that, when a test sample does not truly represent the population (universum) from which it is drawn, the test sample is considered a bias sample. It then becomes virtually impossible to make an accurate statement or to predict about the population.

Experimental Research

This type of research is known in literature by a variety of names. Synonyms are, for instance: the cause and consequence method, before and after design, control group design and the laboratory method. Landman summarises experiential research when he states that it is research designed to study cause and consequence. A clear distinction between the terms experiment and experimental research should be evident. In the former there is normally no question about the interpretation of data in the discovery of new meaning.

Experimental research, however, has control as fundamental characteristic. The selection of control groups, based on proportional selection, forms the basis of this type of research. Experimental research is basically the method that can be applied in a research laboratory. The basic structure of this type of research is elementary: two situations (cause and consequence) are assessed in order to make a comparison. Following this, attempts should be made to treat the one situation (cause) from the outside (external variable) to affect change, and then to reevaluate the two situations. The perceivable changes that occurred can then be presumed as caused by external variables.

Control Group

- *Because*: Control is a fundamental characteristic of this type of research, control groups are a prerequisite. Control groups are selected from a group of selected persons whose experience corresponds with that of the experimental group. The only difference is that they do not receive the same treatment.

- *Variable*: In order to do experimental! research, it is necessary to distinguish clearly between the terms dependent and independent variables. In experimental research it is a prerequisite that the researcher should be able to manipulate the variable and then to assess what the influence of the manipulation on the variable was. A variable is any characteristic (of man or his environment) that can take on different values. Objects are usually not considered as variables-but their characteristics are. As example the following can be considered: a transparency is not a variable (it is an object). The characteristics of the transparency are variables, for example the colour, design etc. In other words, a transparency as an object can take on different values.
- *Independent variable:* According to Landman the independent variable is the circumstances or characteristics which the researcher can manipulate in his effort to determine what their connection with the observed phenomenon is. This means that the researcher has direct control over the variable. As example of an independent variable, is study methods.
- *Dependent variable*: The dependent variable, on the other hand, is the circumstances or characteristics that change, disappear or appear when the researcher implements the independent variable. For example, learning content that should be mastered (student performance) is the dependent variable, while the manipulation of study methods by means of different teaching methods, is the independent variable.
- *Internal and external validity*: The importance of control in conducting experimental research has been pointed

out earlier. A further pre-requisite for this type of research is validity. Validity is a term used in research methodology that indicates the extent to which a test complies with the aim it was designed for.

- *Internal validity*: Internal validity means that the perceived difference in the independent variable (characteristics that change) is a direct result of the manipulation of the obtained research results, and therefore possible to conclude. In experimental design, emphasis is placed on the way in which reference between independent and dependent variables should not be confused by the presence of uncontrolled variables (Landman 1988: 97).
- *External validity*: External validity means that the results of the experimental research should be applied to a similar situation outside the experimental design. The results of the experimental research can then be confirmed in similar situations. (The findings are then considered general.)
- *Ex post facto-research*: Experimental research, where the researcher manipulates the independent variable, whilst the dependable variable are controlled with the aim of establishing the effect of the independent variable on the dependable variable, is also applicable.

The term ex post facto according to Landman is used to refer to an experiment in which the researcher, rather than creating the treatment, examines the effect of a naturally occurring treatment after it has occurred. In other words it is a study that attempts to discover the pre-existing causal conditions between groups. It should, however, be pointed out that the most serious danger of ex post facto-research is the conclusion that because two factors go together, one is the cause and the other is the effect.

Jacobs et al. (1992: 81) refers to the following procedures when conducting ex post facto-research:

- The first step should be to state the problem.

- Following this is the determination of the group to be investigated. Two groups of the population that differ with regard to the variable, should be selected in a proportional manner for the test sample.
- Groups, according to variables, are set equal by means of paring off and statistical techniques of identified independent and dependent variables.
- Data is collected. Techniques like questionnaires, interviews, literature search etc:. are used to determine the differences.
- Next follows the interpretation of the research results. The hypothesis is either confirmed or rejected.

Lastly it should be mentioned that this type of research has shortcomings, and that only partial control is possible.

3

Nano Research Data

DATA COLLECTION

AIM

Data collection is simply how information is gathered. There are various methods of data collection such as personal interviewing, telephone, mail and the Internet. Depending on the survey design, these methods can be used separately or combined. The Data Collection Methodology team (DCM) advises on data collection methods for all ONS social and business surveys and the Census. With clients both within the ONS and the wider government community, we aim to provide expert advice on data collection procedures and carry out research leading to improvements in survey quality.

AIMS

DCM contributes to the development of a common statistical infrastructure for ONS. The team aims to provide business areas with methodological standards and expert advice on procedures to minimise error, in order to produce accurate survey estimates and thereby maintain the quality of data output. DCM is also involved in ongoing research to assess and improve methods of data collection, and in developing new ways of collecting data, for example via the web.

CURRENT AND PLANNED AREAS OF WORK

The main work we do involves giving advice, support and methodological expertise in the following areas:

- Standards and principles of designing, testing,

evaluating and implementing survey processes at the data collection stage

- Expert reviews of questionnaires/data collection instruments
- Cognitive question testing to develop questions for surveys
- Moderation of focus groups to discuss survey topics and concepts
- Measurement and reduction of non-response error
- Implementation of standard response outcome codes
- Methods of calculating response rates
- Methods of identifying key responders for priority response chasing
- Coding and social classification for surveys
- New and improved methods of data collection

DATA EDITING AND IMPUTATION

What is Data Editing and Imputation

Data editing is about detecting and correcting errors in the information returned by the contributors of the data. Imputation is a procedure used for entering a value for a specific data item where the response is missing or unusable.

Aims

The aim is to provide expertise and advice on developing and implementing cost-effective editing and imputation for all ONS data sources, including household and business surveys and administrative sources.

Current and Planned Areas of work

- Provision of support and advice to business areas on editing and imputation procedures
- Production and dissemination of standards and guidance on good practice in editing and imputation
- Exploration of methods and tools to support graphical and output editing
- Identification and evaluation of generalised edit and

imputation tools for possible inclusion in ONS' new statistical infrastructure
- Participation in development of appropriate quality measures associated with editing and imputation processes

DATA COLLECTION ECHNIQUES

OBJECTIVES

At the end of this session you should be able to:
- Describe various data collection techniques and state their uses and limitations.
- Advantageously use a combination of different data collection techniques.
- Identify various sources of bias in data collection and ways of preventing bias.
- Identify ethical issues involved in the implementation of research and ways of ensuring that your research informants or subjects are not harmed by your study.
- Overview of data collection techniques
- The importance of combining different data collection techniques
- Bias in information collection
- Ethical considerations

OVERVIEW OF DATA COLLECTION TECHNIQUES

Data-collection techniques allow us to systematically collect information about our objects of study (people, objects, phenomena) and about the settings in which they occur. In the collection of data we have to be systematic. If data are collected haphazardly, it will be difficult to answer our research questions in a conclusive way.

Example: During a nutrition survey three different weighing scales were used in three villages. The researchers did not record which scales were used in which village. After completion of the survey it was discovered that the scales were not standardised and indicated different weights when weighing the same child.

It was therefore impossible to conclude in which village malnutrition was most prevalent.

Various data collection techniques can be used such as:

- Using available information
- Observing
- Interviewing (face-to-face)
- Administering written questionnaires
- Focus group discussions
- Projective techniques, mapping, scaling

Using Available Information

Usually there is a large amount of data that has already been collected by others, although it may not necessarily have been analysed or published. Locating these sources and retrieving the information is a good starting point in any data collection effort. For example, analysis of the information routinely collected by health facilities can be very useful for identifying problems in certain interventions or in flows of drug supply, or for identifying increases in the incidence of certain diseases. Analysis of health information system data, census data, unpublished reports and publications in archives and libraries or in offices at the various levels of health and health-related services, may be a study in itself. Usually, however, it forms part of a study in which other data collection techniques are also used.

The use of key informants is another important technique to gain access to available information. Key informants could be knowledgeable community leaders or health staff at various levels and one or two informative members of the target group (*e.g.*, adolescents on their sexual behaviour).

They can be involved in various stages of the research, from the statement of the problem to analysis of the data and development of recommendations. Other sources of available data are newspapers and published case histories, *e.g.*, patients suffering from serious diseases, or their relatives, telling their experiences and how they cope.

For example, Noerine Kaleeba *We miss you all: AIDS in the family*. Harare: Women and AIDS Support Network.

Note: In order to retrieve the data from available sources, the researcher will have to design an instrument such as a checklist or compilation sheet. In designing such instruments, it is important to inspect the layout of the source documents from which the data is to be extracted. For health information system (HIS) data, for example, the data compilation sheet should be designed in such a way that the items of data can be transferred in the order in which the items appear in the source document. This will save time and reduce error.

The advantage of using existing data is that collection is inexpensive. However, it is sometimes difficult to gain access to the records or reports required, and the data may not always be complete and precise enough, or too disorganised.

Observing

OBSERVATION is a technique that involves systematically selecting, watching and recording behaviour and characteristics of living beings, objects or phenomena.

Observation of human behaviour is a much-used data collection technique. It can be undertaken in different ways:

- Participant observation: The observer takes part in the situation he or she observes. (For example, a doctor hospitalised with a broken hip, who now observes hospital procedures 'from within'.)

Non-participant Observation

The observer watches the situation, openly or concealed, but does not participate. Observations can be open (*e.g.*, 'shadowing' a health worker with his/her permission during routine activities) orconcealed (*e.g.*, 'mystery clients' trying to obtain antibiotics without medical prescription). They may serve different purposes. Observations can give additional, more accurate information on behaviour of people than interviews or questionnaires. They can also check on the information collected through interviews especially on sensitive topics such as alcohol or drug use, or stigmatising diseases. For example, whether community members share drinks or food with patients suffering from feared diseases (leprosy, TB, AIDS) are essential observations in a study on stigma.

Observations of human behaviour can form part of any type of study, but as they are time consuming they are most often used in small-scale studies. Observations can also be made on objects. For example, the presence or absence of a latrine and its state of cleanliness may be observed. Here observation would be the major research technique.

If observations are made using a defined scale they may be called measurements. Measurements usually require additional tools. For example, in nutritional surveillance we measure weight and height by using weighing scales and a measuring board. We use thermometers for measuring body temperature.

Interviewing

An INTERVIEW is a data-collection technique that involves oral questioning of respondents, either individually or as a group. Answers to the questions posed during an interview can be recorded by writing them down (either during the interview itself or immediately after the interview) or by tape-recording the responses, or by a combination of both. Interviews can be conducted with varying degrees of flexibility.

The two extremes, high and low degree of flexibility, are described below:

- High degree of flexibility:

For example: When studying sensitive issues such as teenage pregnancy and abortions, the investigator may use a list of topics rather than fixed questions. These may, *e.g.*, include how teenagers started sexual intercourse, the responsibility girls and their partners take to prevent pregnancy (if at all), and the actions they take in the event of unwanted pregnancies. The investigator should have an additional list of topics ready when the respondent falls silent, (*e.g.*, when asked about abortion methods used, who made the decision and who paid). The sequence of topics should be determined by the flow of discussion. It is often possible to come back to a topic discussed earlier in a later stage of the interview.

The unstructured or loosely structured method of asking questions can be used for interviewing individuals as well as groups of key informants. A flexible method of interviewing is useful if a researcher has as yet little understanding of the problem or situation he is investigating, or if the topic is sensitive. It is frequently applied in exploratory studies. The instrument used may be called an interview guide or interview schedule.

- *Low degree of flexibility*: Less flexible methods of interviewing are useful when the researcher is relatively knowledgeable about expected answers or when the number of respondents being interviewed is relatively large. Thenquestionnaires may be used with a fixed list of questions in a standard sequence, which have mainly fixed or pre-categorised answers.

For example: After a number of observations on the (hygienic) behaviour of women drawing water at a well and some key informant interviews on the use and maintenance of the wells, one may conduct a larger survey on water use and satisfaction with the quantity and quality of the water.

Though in principle one may speak of loosely structured questionnaires, in practice the term questionnaire appears to be so hooked to tools with pre-categorised answers that we have decided to use the term interview guide for loosely structured tools. However, in reality there is often a mixture of open and pre-categorised answers. In that case we will still use the term questionnaire.

Administering Written Questionnaires

A WRITTEN QUESTIONNAIRE (also referred to as self-administered questionnaire) is a data collection tool in which written questions are presented that are to be answered by the respondents in written form.

A written questionnaire can be administered in different ways, such as by:

- Sending questionnaires by mail with clear

instructions on how to answer the questions and asking for mailed responses;

- Gathering all or part of the respondents in one place at one time, giving oral or written instructions, and letting the respondents fill out the questionnaires; or
- Hand-delivering questionnaires to respondents and collecting them later.

Focus Group Discussions (FGD)

A focus group discussion allows a group of 8-12 informants to freely discuss a certain subject with the guidance of a facilitator or reporter.

Projective Techniques

When a researcher uses projective techniques, (s)he asks an informant to react to some kind of visual or verbal stimulus.

For example: An informant may be provided with a rough outline of the body and be asked to draw her or his perception of the conception or onset of an illness.

Another example of a projective technique is the presentation of a hypothetical question or an incomplete sentence or case/study to an informant ('story with a gap'). A researcher may ask the informant to complete in writing sentences such as:

- If I were to discover that my neighbour had TB, I would . . .;

Mapping and Scaling

Mapping is a valuable technique for visually displaying relationships and resources. In a water supply project, for example, mapping is invaluable. It can be used to present the placement of wells, distance of the homes from the wells, other water systems, etc. It gives researchers a good overview of the physical situation and may help to highlight relationships hitherto unrecognised. Mapping a community is also very useful and often indispensable as a pre-stage to sampling. Scaling is a technique that allows researchers through their respondents to categorise certain variables that they would not be able to rank themselves.

For example, they may ask their informant(s) to bring certain types of herbal medicine and ask them to arrange these into piles according to their usefulness.

The informants would then be asked to explain the logic of their ranking. Mapping and scaling may be used as participatory techniques in rapid appraisals or situation analyses. In a separate volume on participatory action research, more such techniques will be presented. Rapid appraisal techniques and participatory research are approaches often used in health systems research.

Differentiation between data Collection Techniques and data Collection Tools

To avoid confusion in the use of terms, the following table points out the distinction between *techniques* and *tools*applied in data collection.

Table: Data Collection Techniques and Tools

Data Collection Techniques	Data Collection Tools
Using available information	Checklist; Data compilation forms
Observing	Eyes and other senses, pen/paper, watch, scales, microscope, etc.
Interviewing	Interview quide, checklist, questionnaire, tape recorder
Administering written questionnaires	Questionnaire

DIFFERENT DATA COLLECTION TECHNIQUES

When discussing different data collection techniques and their advantages and disadvantages, it becomes clear that they can complement each other. A skilful use of a combination of different techniques can reduce the chance of bias and will give a more comprehensive understanding of the topic under study.

Researchers often use a combination of flexible and less flexible research techniques.

Flexible techniques, such as:

- Loosely structured interviews using open-ended questions,
- Focus group discussions, and
- Participant observation

are also called QUALITATIVE research techniques. They produce qualitative data that is often recorded in narrative form. Qualitative research techniques involve the identification and exploration of a number of often mutually related variables that give INSIGHT in human behaviour (motivations, opinions, attitudes), in the nature and causes of certain problems and in the consequences of the problems for those affected. 'Why', 'What' and 'How' are important questions.

Structured questionnaires that enable the researcher to quantify pre- or post-categorised answers to questions are an example of QUANTITATIVE research techniques. The answers to questions can be counted and expressed numerically. Quantitative research techniques are used to QUANTIFY the size, distribution, and association of certain variables in a study population. 'How many?' 'How often?' and 'How significant?' are important questions. Both qualitative and quantitative research techniques are often used within a single study.

For example: It has been observed in country X that children between 1 and 2-1/2 years, who have already started to eat independently, have unsatisfactory food intake once they fall ill. A study could be designed to address this problem, containing the following stages:

- Focus group discussions (FGDs) with 2 to 5 groups of mothers or in-depth interviews with 10-20 mothers, to find out whether they change the feeding practices for children in this age group when they suffer from (various) illnesses and how mothers deal with children who have no appetite when they are sick (exploratory study);
- A cross-sectional survey, testing the relevant findings of the exploratory study on a larger scale; and

- FGDs with women in the study area to discuss findings and possible questions arising from the survey and to develop possible solutions for problems detected.

In this example, the first, qualitative part of the study would be used to focus the survey on the most relevant issues (mothers' feeding behaviours and reasons for these behaviours) and to help phrase the questions in an optimal way in order to obtain the information that is needed. The second, quantitative part of the study would be used to find out what proportion of the mothers follow various practices and the reasons for their behaviours and whether certain categories of children (*e.g.*, the younger ones or children from specific socio-economic categories) are more at risk than others. The third, qualitative part of the study would provide feedback on the major findings of the survey. Do the conclusions make sense to women in the study area? Have certain aspects been overlooked when interpreting the data? What remedial action is feasible to improve practices related to feeding sick children? It is also common to collect qualitative and quantitative data in a single questionnaire.

Researchers collecting both types of data have to take care that they:

- Do not include too many open-ended questions in large-scale surveys, making data analysis more complicated; and
- Do not use inappropriate statistical tests on quantitative data generated by small-scale studies.

BIAS IN INFORMATION COLLECTION

BIAS in information collection is a distortion in the collected data so that it does not represent reality.

Possible sources of bias during data collection:

Defective Instruments

- *Questionnaires with*:
  - Fixed or closed questions on topics about which little is known (often asking the 'wrong things');

  - Open-ended questions without guidelines on how to ask (or to answer) them;
  - Vaguely phrased questions;
  - 'Leading questions' that cause the respondent to believe one answer would be preferred over another; or
  - Questions placed in an illogical order.
- Weighing scales or other measuring equipment that are not standardised.

These sources of bias can be prevented by carefully planning the data collection process and bypre-testing the data collection tools.

Observer Bias

Observer bias can easily occur when conducting observations or utilising loosely structured group- or individual interviews. There is a risk that the data collector will only see or hear things in which she is interested or will miss information that is critical to the research.

Observation protocols and guidelines for conducting loosely structured interviews should be prepared, and training and practice should be provided to data collectors in using both these tools. Moreover it is highly recommended that data collectors work in pairs when using flexible research techniques and discuss and interpret the data immediately after collecting it. Another possibility-commonly used by anthropologists-is using a tape recorder and transcribing the tape word by word.

Effect of the Interview on the Informant

This is a possible factor in all interview situations. The informant may mistrust the intention of the interview and dodge certain questions or give misleading answers. For example: in a survey on alcoholism you ask school children: 'Does your father sometimes get drunk?' Many will probably deny that he does, even if it is true. Such bias can be reduced by adequately introducing the purpose of the study to informants, by phrasing questions on sensitive issues in a positive way, by taking sufficient time for the interview, and by assuring informants that the data collected will be confidential (Module 10B). It is also important to be careful in the selection of interviewers. In a study soliciting the reasons for the low utilisation of local health services, for example, one should not ask health workers from the health centres concerned to interview the population. Their use as interviewers would certainly influence the results of the study.

Information Bias

Sometimes the information itself has weaknesses. Medical records may have many blanks or be unreadable. This tells something about the quality of the data and has to be recorded. For example, in a TB defaulter study the percentage of defaulters with an incomplete or missing address should be calculated.

Another common information bias is due to gaps in people's memory; this is called *memory* or *recall bias*. A mother may not remember all details of her child's last diarrhoea episode and of the treatment she gave two or three months afterwards.

For such common diseases it is advisable to limit the period of recall, asking, for example, 'Has your child had diarrhoea over the past two weeks?'

Note: All these potential biases will threaten the validity and reliability of your study. By being aware of them it is possible, to a certain extent, to prevent them. If the researcher does not fully succeed, it is important to report honestly in what ways the data may be biased.

ETHICAL CONSIDERATIONS

As we develop our data collection techniques, we need to consider whether our research procedures are likely to cause any physical or emotional harm.

Harm may be caused, for example, by:

- Violating informants' right to privacy by posing sensitive questions or by gaining access to records which may contain personal data;

- Observing the behaviour of informants without their being aware (concealed observation should therefore always be crosschecked or discussed with other researchers with respect to ethical admissibility);
- Allowing personal information to be made public which informants would want to be kept private, and
- Failing to observe/respect certain cultural values, traditions or taboos valued by your informants.

Several methods for dealing with these issues may be recommended:

- Obtaining informed consent before the study or the interview begins;
- Not exploring sensitive issues before a good relationship has been established with the informant;
- Ensuring the confidentiality of the data obtained; and
- Learning enough about the culture of informants to ensure it is respected during the data collection process.

If sensitive questions are asked, for example, about family planning or sexual practices, or about opinions of patients on the health services provided, it may be advisable to omit names and addresses from the questionnaires.

Exercise: Selection of study types and data collection techniques (in plenary) Five health management problems for which studies must be developed are described below.

For each problem you are asked to state:

- What type(s) of study you would propose.
- From whom (or from what) you would collect the data required for each study (your study populations).
- For each study population: which data collection techniques you would use.
- You suspect that a large proportion of women in your region (pop. 1,000,000) is anaemic, in particular women at childbearing age. You would like to determine how big the problem is, and whether women perceive it as a problem. Furthermore you would like to know whether and how women themselves could contribute to improving their anaemic condition.

- A district health team evaluated its malaria spraying programme by looking at available records and reports. It did not find significant flaws in the functioning of the services in different divisions and villages. Nevertheless, the incidence of malaria and mosquito counts show peaks in certain villages that the DHT cannot explain. It wants to find out if there is something wrong with the services.
- A community survey in your region (pop. 2,000,000) indicates that 12% of the adults (15–60) in the capital are HIV positive, as compared to 7% in road side settlements and 2,5% in the rural area (where 80% of the population live). You want to step-by-step introduce an intensive STD programme in all 11 hospitals and 180 health centres, hoping to decrease the HIV incidence. You would like to evaluate whether the STD programme has an effect. Apart from the three main questions stated above: How would you organise the evaluation time-wise? Are any ethics involved? What could be two important biases in the study?
- You are a midwife who has just been appointed as head of a maternity unit in a district hospital. You suspect that the number of low birth-weight babies is increasing, and you would like to know more about the physical and socio-economic conditions of the mothers to see if remedial actions should be taken. The clinic records at present are not complete enough to draw conclusions and you have neither the time nor the money to do a district-wide community survey.
- There are long queues (waiting times), at the out-patient department of your district hospital. You are concerned about this and you would like to find out to what extent the problem may be related to the organisation and management of the department and whether certain bottlenecks can be identified. In a later stage of the research you would like to try to

eliminate some of the bottlenecks and see whether there is improvement.

- *Group work*: Selection of study type and data collection techniques
- Decide what type(s) of study you will apply in your own research proposal.
  - Make your choice on the basis of your research objectives and the variables you would like to include in the study. (Hang objective and variables on a wall or flip chart board so the whole group can see them during the current group work session.)
- Determine what data-collection techniques you will use for each variable in your study.
  - Display the table that was prepared during the group work on selection of variables. For each variable determine the source of data and method(s) of data collection.
  - For some variables it may be necessary to collect additional data in order to be able to define the variable and determine the scale of measurement.

Example:	Factor:	Inadequate knowledge of patients about TB treatment
	Variable:	Knowledge of patients
	Scale of measurement:	Percentage or number of items of advice that are recounted by patients

- To determine the 'items' that should be used in the variable, it may be necessary to do a focus group discussion (FGD) with the staff treating TB patients and/or with a panel of experts on TB. Subsequently the results of the FGD could be used to develop questionnaires for interviews with patients. Hence the methods of data collection would be a FGD (to develop knowledge indicators) and face-to-face interviews with patients.
- Summarise which data collection techniques you will use and which groups or records will form the sources of your data for each tool.

- Decide whether there are any ethical problems with the type of study or data collection tools you propose.

Trainer's Notes

Module 10A

Table. Timing and teaching methods

1 hour	Introduction to data collection techniques and discussion
1 hour	Exercise: Selection of study types and data collection techniques
1 hour	Group work: Selection of study types and data collection techniques
1 hour	Plenary
4 hours	Total time

Introduction and discussion

- Present an overview of the various data collection techniques. Give examples from the participants' fields of interest.
- If it would be useful for the participants' research projects, you can introduce additional research techniques using the material in Module 10C (Focus group discussions) or other sources, *e.g.*, the volume on Participatory Action Research to be published by AFRO in the HSR series or Pretty *et al.* Participatory Learning and Action.
- Explain the difference between data collection techniques and data collection tools.
- Let the participants mention possible advantages and disadvantages of the various data collection techniques.
- Explain at what times qualitative research techniques are most useful and when quantitative techniques are more appropriate. Make sure that participants understand the advantages of combining quantitative and qualitative research techniques, preferably giving examples from one or more of the projects they are developing.
- Identify different possibilities for bias, using examples from the participants' own studies.

- Let the groups come up with examples of ethical issues that might play a role in their studies.

Exercise: Selection of study types and data collection techniques

- This exercise is designed to give participants some experience in choosing types of studies appropriate for typical situations before they have to select types of studies for their own proposals.
- Stress that objectives, if well formulated, should help to determine the appropriate study type(s).
- Ask participants to divide into sub-groups of 4-5 persons to do the exercise. Each sub-group should be assigned 2 topics. Allow 30 minutes for the groups to complete their work. Let them write the answers on an overhead transparency or flipchart.
- In the plenary (30 minutes) ask each group to answer the questions posed for one topic. Let other groups who discussed the same topic comment and add their own suggestions.

Group work: Selection of study type and data collection techniques

- This group work session is an important one, as it combines both the selection of study type and choice of data collection techniques. Work closely with your group, helping them to go step by step through the process described for the group work.
- Make sure they understand why it is important to refer back to their variables and the table they developed during Module 8, as they choose the data collection techniques they need.

The delay of two years would form an *ethical problem* as soon as the preliminary results would indicate that the extra input in STD treatment in the intervention HCs indeed reduces the risk of HIV infection. If that is the case, experiments are sometimes stopped in order to allow the control group to profit from the intervention as well. One important *bias* could be that STD patients in control villages would have heard about the available treatment for STD in the experimental HCs and go there for treatment. To be able to control this potential bias, the research team analysed the treatment registers in the experimental HCs for the location of all STD patients who received treatment and deducted patients from the control villages (which appeared to be exceptions).

Another bias may be caused by intensified health education on STD symptoms and treatment over radio and TV, which could influence the treatment seeking behaviour of members of the experimental as well as of the control group.

SAMPLING PROCEDURE

For most of our national surveys of the general public, we conduct telephone surveys using a random digit sample of landline and cell phone numbers in the continental United States. Some of these general population surveys include additional, larger samples of subgroups, such as African Americans or young people. (These are called "oversamples.") We also conduct surveys of people in particular states or regions, where our sample is limited to residents of these areas, and international surveys that involve sampling people in multiple countries. Lastly, we occasionally conduct surveys of special populations, such as foreign policy officials or journalists. The importance of probably sampling is discussed more broadly in the final section "Why probability sampling."

RANDOM DIGIT DIALING

The typical Pew Research Center for the People and the Press national survey selects a random digit sample of both landline and cell phone numbers in the continental United States. As the proportion of Americans who rely solely or mostly on cell phones for their telephone service continues to grow, sampling both landline and cell phone numbers helps to ensure that our surveys represent all adults who have access to either. We sample landline and cell phone numbers to yield a ratio of approximately three landline interviews to each cell phone interview. This ratio is based on an analysis that attempts to balance cost and fieldwork considerations as well as to improve the overall demographic composition of the sample (in terms of age, race/ethnicity and education). This ratio also ensures a minimum number of cell only respondents in each survey.

The design of the landline sample ensures representation of both listed and unlisted numbers (including those not yet listed) by using random digit dialing. This method uses random generation of the last two digits of telephone numbers selected on the basis of the area code, telephone exchange, and bank number. A bank is defined as 100 contiguous telephone numbers, for example 800-555-1200 to 800-555-1299. The telephone exchanges are selected to be proportionally stratified by county and by telephone exchange within the county. That is, the number of telephone numbers randomly sampled from within a given county is proportional to that county's share of telephone numbers in the U.S. Only banks of telephone numbers containing three or more listed residential numbers are selected.

The cell phone sample is drawn through systematic sampling from dedicated wireless banks of 100 contiguous numbers and shared service banks with no directory-listed landline numbers (to ensure that the cell phone sample does not include banks that are also included in the landline sample). The sample is designed to be representative both geographically and by large and small wireless carriers.

Both the landline and cell samples are released for interviewing in replicates, which are small random samples of the larger sample. Using replicates to control the release of telephone numbers ensures that the complete call procedures are followed for the entire sample. The use of replicates also ensures that the regional distribution of numbers called is appropriate. This also works to increase the representativeness of the sample. When interviewers reach someone on a landline phone, they ask to speak with "the youngest male, 18 years of age or older, who is now at home." If there is no eligible male at home, interviewers ask to speak with "the youngest female, 18 years of age or older, who is now at home." This method of selecting respondents within each household improves participation among young people who are often more difficult to interview than older people because of their lifestyles. Unlike a landline phone, a cell phone is assumed in Pew Research polls to be a personal device.

CELL PHONES

As the proportion of Americans who rely solely or mostly on a cell phone for their telephone services now approaches 20%, the representativeness of telephone surveys based on a random sample of households with landline telephones has come under increased scrutiny. More pollsters and survey methodologists are researching the influence of cell phones on telephone surveying.

Public Opinion Quarterly dedicated a special issue to the topic of cell phones in 2007 Cell Phone Numbers and Telephone Surveying in the U.S. The Pew Research Center for the People and the Press decided to call cell phones for our Fall 2008 election surveys and is now including a cell phone sample in nearly all polls.

One of the main challenges of surveying cell phone users is drawing a representative sample of this group. Drawing samples for all telephone surveys is now more complicated because of the introduction of cell phone numbers and number portability (*i.e.*, where people can keep their numbers when they move or change service providers or port a landline number to a cell phone).

Telephone numbers are assigned different prefixes, which can be used to identify whether the number is for a landline or cell phone, but there are also mixed or shared prefixes that include both landline and cell numbers. In addition, people who forward their calls (*e.g.*, from their landline number at home or work to their cell) may appear as a landline number even when they are actually talking on their cell phones.

Most telephone surveys use the household as the sampling unit because landline telephone numbers have typically been shared among all members living in a household. Once a sample of landline telephone numbers is drawn, a separate selection procedure is used to give all adults living in a given household a chance of selection (such as asking for the youngest adult male or female). However, the situation is more complicated for cell phone users because cell phones are often considered individual rather than shared devices, so the person who answers the phone usually becomes the respondent, whether he or she is the primary user of the phone or shares the cell phone with others.

Although some surveyors have experimented with selecting among the users of a shared cell phone, it is still uncertain whether the benefits of this approach outweigh the disadvantages, such as potentially lower response rates. In addition, many people under the age of 18 (and thus not eligible for most national surveys) have cell phones. Substantial time and costs are incurred screening out these ineligible respondents.

Several additional issues arise when identifying the geographic location of a cell phone number. The geographic information that can be derived from cell phone numbers is much less precise than for landline telephone numbers. The boundaries of wireless service areas are often larger than landline service areas. The geographic information is based on the rate center where the phone was purchased, rather than where the person lives.

And many people move without changing their cell phone numbers. Based on a comparison of geographic information provided with the sample to that derived from respondents' self-reported zip codes, as many as 10% of respondents live in a different state and nearly 40% in a different county than the sample would predict. This issue is of particular concern for sampling cell phones within a geographic area.

Although respondents who do not live in the area may be identified by a screener question, people who do live in the area—but have cell phones from elsewhere-are likely to be excluded from the survey. In addition, cell phone penetration rates are not available for most counties or states. Because of this, there is no way to select cell phone numbers proportional to their size within these geographic areas.

In addition to the different procedures necessary for sampling cell phone numbers, there are also substantial challenges with interviewing people on their cell phones. Those challenges are discussed in more detail in Cell phone surveys.

OVERSAMPLES

For some surveys, it is important to ensure that there are enough members of a certain subgroup in the population so that more reliable estimates can be reported for that group. To do this, we oversample members of the subgroup by selecting more people from this group than would typically be done if everyone in the sample had an equal chance of being selected.

Because the margin of sampling error is related to the size of the sample, increasing the sample size for a particular subgroup through the use of oversampling allows for estimates to be made with a smaller margin of error. A survey that includes an oversample also typically weights the results so that members in the oversampled group are weighted to their actual proportion in the population; this allows for the overall survey results to represent both the national population and the oversampled subgroup.

For example, African Americans make up about 11% of the total U.S. population, according to the U.S Census. A survey with a sample size of 1,000 would only include approximately 110 African Americans. The margin of sampling error for African Americans then would be around 10 percentage points. Estimates about African Americans from this survey thus could fall within a 20 point range, which is often too imprecise for many detailed analyses that surveyors want to perform.

In contrast, oversampling African Americans so that there are roughly 450 interviews completed with people in this group reduces the margin of sampling error to about 5 percentage points and improves the reliability of estimates that can be made. Unless a listed sample is available or people can be selected from prior surveys, oversampling a particular group usually involves incurring the additional costs associated with screening for eligible respondents.

An alternative to oversampling certain groups is to increase the overall sample size for the survey. This option is especially desirable if there are multiple groups of interest that would need to be oversampled. However, this approach often increases costs because the overall number of completed interviews needs to be increased substantially to improve the representation of the subgroup(s) of interest.

The studies reported below included an oversample in the survey design. Among other groups, these studies included oversamples of young people, African Americans, internet users, and parents of young children. See the "About the Survey" section in each report for details.

REGIONAL

Many surveys conducted in the U.S. are not national in scope but instead are designed to represent residents of a single community, city, county, state, or region. These surveys tend to have similar sampling procedures to many national surveys, but the people sampled are limited to the geographic area of interest. For example, residents of states can be sampled using the random digit dialing procedures described for our national surveys. However, there may be considerable error with sampling cell phones in a particular geographic area as discussed in the section on sampling cell phones.

The Pew Research Center has conducted a number of state surveys, especially in the context of upcoming presidential primaries. And, we have conducted some special surveys of metropolitan area residents of Philadelphia, New York and Washington.

INTERNATIONAL

International polling poses special challenges to surveyors. The sampling procedures are specific to each country and may affect what mode is selected for data collection. In addition, for multi-country studies data collection needs to be coordinated across multiple countries and data collection organizations. One of the most important aspects of international surveying is the attention given to how concepts and questions are interpreted across multiple cultures and languages.

Most of the international surveys conducted by the Pew Research Center are through the Pew Global Attitudes Project. The Pew Global Attitudes surveys were launched on a regular basis in 2002. The Project has conducted a series of worldwide public opinion surveys that encompasses a broad array of subjects, ranging from people's assessments of their own lives to their views about the current state of the world and important issues of the day. Depending on the mode best suited to a country, the surveys are conducted by telephone or face-to-face. Most of the national surveys are representative of the entire population of a given country though at times data collection is limited in some way-typically to more urban areas-and results in a survey that is not fully representative nationally.

ELITES AND OTHER SPECIAL POPULATIONS

Representative surveys can be conducted with almost any population imaginable. It is common for surveyors to want to collect information from experts or elites in particular fields (such as policy-makers, elected officials, scientists or news editors) and other special populations (such as special interest groups, people working in particular sectors, etc.). The principles of drawing a representative sample are the same whether the sample is of the general population or some other group. Decisions must be made about the size of the sample and the level of precision desired so that the survey can provide accurate estimates for the population of interest, and any subgroups within the population that will be analysed.

Some special challenges arise when sampling these populations. In particular, it may be difficult to find a sampling frame or list for the population of interest and this may influence how the population is defined. In addition, information may be available for only some ways of contacting potential respondents (*e.g.*, e-mail addresses but not phone numbers) and may vary for people within the sample. If most members in the population of interest have internet access and e-mail addresses are available, the internet often provides a convenient and inexpensive way to survey experts or other special populations.

The Pew Research Center occasionally conducts surveys of opinion leaders, especially those in public policy roles. The opinion of elites is often compared with that of the general public to better determine whether these groups have similar or different opinions. The Pew Research Center has conducted several surveys designed to be representative of a special population including surveying journalists, Muslim Americans, Howard Dean's campaign supporters during the 2004 presidential primary campaign, political campaign consultants and constituent groups from a sample of federal agencies.

SAMPLE SIZE

THE SAMPLE SIZE

The level of precision needed for the survey estimates will impact the sample size. However, it is not as easy to determine the sample size as one may think. Generally, the actual sample size of a survey is a compromise between the level of precision to be achieved, the survey budget and any other operational constraints, such as budget and time.

In order to achieve a certain level of precision, the sample size will depend, among other things, on the following factors:

- *The variability of the characteristics being observed*: If every person in a population had the same salary, then a sample of one person would be all you would need to estimate the average salary of the population. If the salaries are very different, then you would need a bigger sample in order to produce a reliable estimate.
- *The population size*: To a certain extent, the bigger the population, the bigger the sample needed. But once you reach a certain level, an increase in population

no longer affects the sample size. For instance, the necessary sample size to achieve a certain level of precision will be about the same for a population of one million as for a population twice that size.

- *The sampling and estimation methods*: Not all sampling and estimation methods have the same level of efficiency. You will need a bigger sample if your method is not the most efficient. But because of operational constraints and the unavailability of an adequate frame, you cannot always use the most efficient technique.

WHY PROBABILITY SAMPLING

Probability sampling, where a small randomly selected sample of the population can be used to estimate the distribution of an attitude or opinion in the entire population with statistical confidence, provides the foundation for survey research and political polling. The basis of probability-based random sampling is that every member of the population must have a known, non-zero chance of being selected. Probability sampling provides the means by which the margin of sampling error can be calculated and the level of confidence in survey estimates reported. Sampling error results from collecting data from some rather than all members of the population and is highly dependent on the size of the sample.

We report a margin of sampling error for the total sample for each survey and sometimes for key subgroups (*e.g.*, registered voters, Democrats, Republicans, etc.). For example, the sampling error for a typical Pew Research Center for the People and the Press national survey of 1500 completed interviews is plus or minus 3 percentage points with a 95% confidence interval. This means that in 95 out of every 100 samples of the same size and type, the results we would obtain will vary by no more than plus or minus 3 percentage points from the result we would get if we could interview every member of the population. Thus, the chances are very high (95 out of 100) that any sample we draw will be within 3 points of the true population value.

NONPROBABILITY-BASED SAMPLING

For some surveys, it may not be feasible or practical to draw a random probability sample of the population. In these cases, various types of nonprobability-based samples may be used, such as convenience, volunteer, purposive, and quota samples. Nonprobability samples are often used for qualitative or exploratory research, such as focus groups or in-depth interviews. Many internet panel surveys, which have increased in recent years, rely on volunteer samples. Some internet panels, however, have been developed to randomly recruit people using another mode so that random probability sampling methods can be used. For nonprobability-based samples, the relationship between the sample and the population is unknown. That means there is no theoretical basis for computing or reporting a margin of sampling error and thus for knowing how representative the sample is of the population as a whole.

A small proportion of Americans (about 2%) do not have access to any phone. In practice this assumption is not always correct, as some people share cell phones. But it is still uncertain whether the benefits of sampling among the users of a shared cell phone outweigh the disadvantages, such as lower response rates.

SAMPLE MEMBER

SELECTION

Sampling allows statisticians to draw conclusions about a whole by examining a part. It enables us to estimate characteristics of a population by directly observing a portion of the entire population. Researchers are not interested in the sample itself, but in what can be learned from the survey—and how this information can be applied to the entire population.

It is essential that a sample survey be correctly defined and organized. If the wrong questions are posed to the wrong people, statisticians will not receive information that will be useful when applied to the entire population. In the context of a national statistical agency like Statistics Canada, the following steps are needed to select a sample and ensure that this sample will fulfill its goals.

Establish the Survey's Objectives

The first step in planning a useful and efficient survey is to specify the objectives with as much detail as possible. Without objectives, the survey is unlikely to generate usable results. Clarifying the aims of the survey is critical to its ultimate success. The initial users and uses of the data should be identified at this stage.

The pros and cons of a census versus a sample survey or the use of administrative records should be evaluated and a decision made as to the most appropriate method. (At this point, we will assume that a sample survey is the best way to proceed in order to obtain the information we need. This assumption will hold true for the remainder of the sample selection steps, even though many of the steps mentioned will also apply to the other methods.)

Define the Target Population

The target population is the total population for which the information is required. For example, if you were to conduct a survey about the most popular types of cars in Saskatchewan, then the target population would be every car in Saskatchewan. The units that make up the population must be described in terms of characteristics that clearly identify them.

Specifically, the target population is defined by the following characteristics:

- *Nature of data required*: About persons, hospitals, schools, etc.
- *Geographic location*: The geographic boundaries of the population have to be determined, as well as the level of geographic detail required for the survey estimate (by province, by city, etc.).
- *Reference period*: The time period covered by the survey.

- Other characteristics, such as socio-demographic characteristics (interest in a particular age group, for example) or type of industry.

Decide on the Data to be Collected

he data requirements of the survey must be established. To ensure that the requirements are operationally sound, the necessary data terms and definitions also need to be determined.

Set the Level of Precision

As mentioned in the section on Sampling error, there is a level of uncertainty associated with estimates coming from a sample. For example, if you are trying to estimate the average distance between home and school for students in your class of 25 from a sample of 5 persons, your estimate will depend on who the 5 sampled students are. If the 5 sampled students also live close to the school, the results will not be able to represent the class accurately. This sample-to-sample variation is what causes the sampling error. Statisticians can estimate the sampling error associated with a particular sampling plan, and try to minimize it.

When designing a survey, the acceptable level of uncertainty in the survey estimates has to be established. This level depends on what the end use of the results will be and on the size of the overall budget. The bigger the budget, the more resources available, and thus, less chance for error. And if the end result is to serve a specific purpose, then the acceptable level of uncertainty would be smaller than an end result that is simply looking for general trends.

THE SAMPLE DESIGN

Once the objectives, guidelines and definitions have been worked out, the statistician can work on the survey plan.

The survey plan is divided into three parts:

- *Sample design*: How the sample will be collected.
- *Estimation techniques*: How the results from the sample will be extended to the whole population.

- *Measures of precision*: How the sampling error will be measured.

The estimation techniques and measures of precision are discussed in a later section. For the moment, we will look at the sample design.

The following steps lead to the complete determination of the sample design:

- Determine what the survey population will be (*e.g.*, students, men aged 20 to 35, newborn babies, etc.).
- Choose the most appropriate survey time frame.
- Define the survey units.
- Establish the sample size (*e.g.*, a sample of 100 from a population of 1,000).
- Select a sampling method.

THE SURVEY POPULATION

The target population must be defined early in the survey-designing process. This is the population for which information is required. However, some members of the population have to be excluded because of operational constraints: the high cost of collecting data in some remote areas, the difficulty of identifying and contacting certain components of the target population, etc. For example, because it would be too difficult to locate and survey each car owned by every resident in Saskatchewan, a survey population of just the major cities and towns might be conducted instead. When some of the members of the target population are excluded, we call the included population the survey population or, what is sometimes called, an *observed population*. The target population is the population we want to observe while the survey population is the population we can observe.

The goal of this process is to have the survey population as close as possible to the target population. It is also very important that the users of the data be informed of the differences between the two populations, as the results of the survey will apply only to the survey population. For example, a target population for a survey could be all Canadians aged

15 years and over (on a particular reference date), while the survey population could exclude residents of the Yukon, Nunavut and Northwest Territories, persons living on Aboriginal reserves, full-time members of the Canadian Armed Forces and residents of institutions.

These Canadians might be excluded for various reasons: to survey people in the territories might prove to be difficult and expensive, military personnel may not be available for surveying if they are out on a mission, etc. Using this example, about 2% of the target population would be excluded from the survey population.

THE SURVEY FRAME

The survey frame, also called the sampling frame, is the tool used to gain access to the population. There are two types of frames: list frames and area frames. A list frame is just a list of names and addresses that provide direct access to 'individuals' (*e.g.*, a list of hospitals, a list of restaurants, a list of students at a university). Area frames are a list of geographic areas that provide indirect access to individuals (*e.g.*, the neighbourhoods in a city). This type of access is called indirect because first, a list of geographic areas must be selected and then, access to individuals within each selected area must be worked out.

For instance, suppose that you were surveying a rural town in Quebec to see what percentage of residents are farmers. If you were provided with an area frame, then you would be able to locate which roads to visit, but you would still have to find out the names and addresses of the residents on each road.

When there is no single frame that is appropriate, multiple frames can be used. Some sampling techniques using both types of frames will be discussed later. A good frame should be complete and up-to-date; no member of the survey population should be excluded from the frame or duplicated on the frame (represented more than once); and no unit that is not part of the population (*e.g.*, deceased persons) should be on the frame.

The frame chosen will impact the selected survey population. For instance, if a list of telephone numbers is used to select a sample of households, then all households without telephones are excluded from the survey population.

THE SURVEY UNITS

There are three types of units that have to be accurately identified in order to avoid problems during the selection, data collection and data analysis stages.

They are as follows:

- The sampling unit is part of the frame and therefore subject to being selected.
- The respondent unit or reporting unit provides the information needed by the survey.
- The unit of reference or unit of analysis—the unit about which information is provided—is used to analyse the survey results.

For example, in a survey about newborns in Edmonton, the sampling unit might be a household, the reporting unit one of the parents or a legal guardian, and the unit of reference the baby. The sampling units may differ depending on the frame used. This is why the survey population, survey frame and survey units are defined in conjunction with one another.

THE SAMPLING METHOD

There are two types of sampling methods: probability sampling and non-probability sampling. The difference between them is that in probability sampling, every unit has a 'chance' of being selected, and that chance can be quantified. This is not true for non-probability sampling; every item in a population does not have an equal chance of being selected. The next section will describe features of both types of sampling and detail some of the methods related to each type.

DATA AND METHODS OF COLLECTING DATA

MEASUREMENT OF DISABILITY

Collecting data about persons with disabilities and their

lives is difficult. It poses unique problems that data collection developers need to address in the design phase of the collection process. Two issues need to be addressed at the outset: what kind of data collection instrument should be used, and what unit of measurement should be employed.

Collection Instruments

The main types of instruments for collecting data about persons with disabilities are:

- Population censuses
- Sample surveys (either general social surveys or specific health and disability surveys)
- Administrative collections and registries

Each of these tools can be used to measure aspects of disability in a population and each has its strengths and weaknesses.

The Population Census certainly has the advantage of providing complete population coverage. Unfortunately, it is difficult to collect accurate information about disability in a census since time constraints make it unlikely that more than 4 to 6 general disability questions can be asked. Censuses, in some instances, also undercount children with disabilities and people with mild or moderate disabilities (in cases where the response categories are limited to 'Yes or No' options only). Still, for a crude measure of disability, and in the absence of other collection instruments, the census is useful.

Sample surveys are shorter surveys designed to be administered to a sub-population selected by some other instrument (often a census) that focus on specific issues. They are often put into the field to answer specific questions about a population. As such, they provide the opportunity to ask more detailed questions about disability. More detailed information is useful in itself, of course, but it also helps to reduce the number of false positive and negative responses, thereby offering a more accurate prevalence measure. A sample survey may be an independent survey focusing entirely on disability, or a disability module added to an existing survey.

Administrative collections and registers are composed of data that is collected as part of the normal operation of some service or programme. An example is the information found on a client intake form. These collections provide useful information on the characteristics of people accessing disability services as well as details about the services provided. They cannot give an accurate measure of disability prevalence since there is no guarantee of coverage and they tend to incorporate double counting. The quality of administrative register data is closely related to the quality of the administrative system, in particular, how well it has been maintained and how closely the concepts align with the disability concepts of interest.

These three instruments for gathering disability information are discussed in detail in the rest of this chapter.

Choice of Selection and Measurement unit

The second preliminary issue that needs to be addressed, whatever data collection instrument is chosen, is how to select the unit for which disability is to be measured. If the selection unit is the individual, then the individual will also be the measurement unit; if the selection unit is a collection of people-invariably a household-then, a decision has to be made as to whether the measurement unit is the household itself (that is, all individuals in the household) or some individual in the household that meets specified criteria (age, gender, unemployed, and/or others).

These decisions depend in part on what kinds of data are needed. Is data required for the number of persons with disabilities and their characteristics, or for an estimate on the number of households that include individuals with disabilities? Data about individuals is important, but estimates at the household level are also useful for detailed analysis of living arrangements and access to help and assistance.

There are also issues of practicality and efficiency. Having the household as the selection unit means that the survey or interview can be conducted in a setting in which information about other people can be efficiently collected, even if only one member of the household is given the full interview. Questions asked of a single household member may reveal another individual with a disability, thereby increasing the sample of person with disability, without adding to the number of households screened.

A note of caution applies, however, if the private household is chosen as the measuring unit. In such a case, thought needs to be given to the fact that this choice excludes residential care units, retirement homes, and other institutions such as prisons. Excluding people living in institutions underestimates the levels of disability for older people and for those with particular types of disability such as psychiatric disabilities. For a complete picture, if feasible, household surveys should be supplemented with institutional surveys of disability.

POPULATION CENSUSES

Many countries have collected information about disability in their national population censuses. In countries without a good household survey program, the census may be the only possible source for estimating disability prevalence and gaining an approximate estimate of types of disability in the country.

The amount of information on disability that can be collected in the population census is very limited, and is often confined to a single question. With only one question, false positive and false negative responses are more common and no complete measure of the number of persons with disabilities is possible, especially among children and the elderly. Still, census data should be readily used, where available, to develop more detailed follow-up surveys *for a discussion of the Canadian experience with a post-census survey on disability*).

Many countries use both short and long forms of census questionnaires The short form is for complete coverage of core topics, such as age, sex and location, and may also have a question on disability. A probability sample is then selected for the long form to be administered. The long form includes all questions in the short form plus supplementary questions for more detailed coverage of selected topics. Questions more detailed than a single general question on disability may be included in the long form.

In response to an increasing concern for more information on the population of persons with disabilities, the Statistics and Census Service of Macau included questions on disability in its 2001 Census. The short form, which was administered to 80 per cent of households, included one generic question: "Is anyone in your household physically or mentally disabled?". In the long form questionnaire, completed by 20 per cent of households, the same disability identification question was asked followed by a question about the respondent's primary disability, and a second question asking whether they had ever used any disability services.

The resulting estimate of the number of person with disabilities from the census was 1.3 per cent of the resident population.

SAMPLE SURVEYS

Sample surveys are good methods of collecting data about persons with disabilities. They use sample selection procedures to identify a sub-population that is representative of the total population, unlike censuses that are designed to enumerate every household or individual in the country.

While sample surveys are considerably cheaper than censuses, the size of the sample affects the quality of the data, which is particularly subject to sampling error (refer to Chapter 5 for the detailed discussion of sampling error). This can frustrate analyses that rely on cross-tabulation.

A small sample size also means that sample surveys cannot provide detailed information for very small geographical areas. Where such data is needed, census or administrative data collections or sophisticated data modelling techniques are more useful options.

Determining sample sizes required to produce statistics with an acceptable level of sampling error is the job of experienced survey methodologists. Unfortunately, the technical details of sample designs are beyond the scope of this manual. The UN Guidelines and Principles for the Development of Disability Statistics has some helpful comments on sample size for disability surveys. (As a rough guide, surveys that have sufficient size to yield valid unemployment estimates for a given geographic area are usually large enough to produce disability estimates for the same area for as long as the prevalence rates are similar.)

As mentioned, a sample survey for disability can either be an independent or dedicated survey, or a module to other surveys, such as a national health survey, a general social survey or labour force survey. Where these general social surveys are cyclical, it is very useful to have a disability module or a small set of disability questions added on.

As Dedicated Disability Surveys

Surveys specifically dedicated to collect disability data are good sources of information about prevalence rates, the causes and types of disability, underlying health conditions, severity and duration of disability, and the use of and need for assistive devices, changes in environment, policies and public awareness on disability.

Dedicated disability surveys maximise the amount of information that can be collected to meet users' needs. Many countries have not attempted to conduct such disability surveys because they are more costly than simply adding a few questions into a census or already existing sample surveys. An example of a country that conducts a dedicated disability survey is provided.

The Survey of Disability, Ageing and Carers provides data on disability prevalence, need for support, and characteristics of persons with disabilities, older people and those who provide care. Information is collected from private households and cared accommodations (hospitals, nursing homes, aged care and disability hostels and other homes such as children's homes). The survey uses computer-assisted personal interviews. Since 1981, it has been conducted every five years.

Sample size: The sample size is approximately 14,000 private dwellings and 300 non-private dwelling units. The carer sample is approximately 550 dwellings. The final sample depends on the number of people in each household or non-private dwelling, and comprises approximately 36 000 people for the household component and 5 000 people for the cared accommodation component.

Data collected:

- Household composition
- Demographic information about all household members
- People selected for personal follow-up interview (people with long-term health conditions or with a disability, who are aged 60 years or over, or someone who regularly provides informal care to someone with a disability)
  - Socio-economic characteristics (education, labour force participation, income, and housing)
  - Impairments, long term health conditions associated with main disability condition
  - Difficulties experienced and help required for activities such as self care and mobility
  - Types of assistance received for a range of activities, met and unmet needs for assistance

Use of aids and equipment

As a Module to other Sample Surveys

In many countries, national household surveys-covering topics such as health, education, living conditions, family income and expenditure, labour force participation, nutrition, time use, crime, and so on-are conducted on a regular basis. Whether every five years, annually, quarterly or monthly, these surveys try to identify short and long-term social trends.

A cost-efficient and effective way of collecting disability data is to add disability questions-or a 'disability module' - to one or several of these cyclical surveys. This approach has been used in many countries. Most commonly, disability modules have been added to labour force or health surveys, and sometimes to living conditions surveys. A disability module was added to Indonesia 's 2003 Socio Economy Survey. The disability population was identified using the question, "Does he/she have a disability?" with a Yes/No response. Where positive responses were received to the disability question, codes for 'type of disability' (7 codes for mostly severe impairments) and 'main cause of disability' (5 codes) were entered into the person's matrix.

This form of questions is expected to yield low disability prevalence rates because the survey relies on self-identification of disability and the only types of disability asked about are severe impairments.

The general approach to including a disability module on a non-disability survey will be useful for the Indonesian government, because disability data can be combined with the other social and economic data collected to analyse the experiences of persons with disabilities (or, at least, those with selected severe impairments).

There are advantages to on-going sample surveys rather than ad hoc or one-off sample surveys for disability. On-going surveys can make maximum use of the resources initially expended, including the time and other resources used in preparing for the data collection, as well as the trained personnel and other resources dedicated to collecting, processing, and analyzing the data on a regular basis.

On-going survey programs also offer opportunities to learn from previous experiences so that the quality and usefulness of the information produced can be improved. They allow for measurement of change over time in key indicators such as frequency of types of disability, severity of disability, quality of life, opportunities and participation of persons with disabilities, and rehabilitation needs. These data can be exceedingly valuable for policy development and evaluation.

ADMINISTRATIVE COLLECTIONS

What is an Administrative Data Collection

Besides surveys and censuses, administrative data

collection (general purposeadministrative systems or administrative registries) intended to serve persons with disabilities can be an important method of gathering information about these people and their characteristics. In this method, any information collected is organized and becomes part of normal service administration procedure such as the information gathered using standard client intake forms for health, rehabilitation, or social work services. Administrative data collections can take several forms, depending on the nature of the service, the format used, the type of information collected, and the method and frequency of collection. Some examples are population registers, vital registration systems, social security systems, registries of occupational injuries, rehabilitation programmes, assistive device services, and other services specifically designed to serve the needs of persons with disabilities.

Administrative registries are databases of records of people with particular characteristics and set up as part of the administration of services to these individuals. Registers are either established during a registration survey or other point in time or they can be on-going and regularly updated.

Administrative records and registers, moreover, often provide unique information about persons with disabilities. The information is usually collected for reasons concerning the administration of the programme or service, but with care, can prove to be useful. For example, data about children and adolescents in special education programmes is an indication of participation rates in education; general invalid pension registries are often organized by disease or impairment; and domestic care allowances help to track rates of children and adults with severe disabilities. Often these data are collected annually, which provides a source for trend data on the prevalence of impairments or disabilities.

Ministries, government departments, advocacy groups, and service providers often maintain administrative records which they use to monitor and evaluate programs and services. Given confidentiality concerns, only aggregated data may be available, but this can still be useful.

It must be kept in mind, however, that data from registration systems cannot provide information about persons with disabilities who need a service or programme but do not receive it. Data about unmet need has to come from other sources. Since the information collected in administrative data collections is limited to people receiving services, or otherwise known to the service agency (as in the case of people on a waiting list), these collections have limited coverage. Therefore administrative collections, despite all their benefits, are not good sources for estimating overall disability prevalence.

What are the Benefits

As compared with censuses and sample surveys, administrative data collections have some advantages for disability data users: Data from administrative collections are generally available on a regular or on-going basis, as they are part of regularly updated information systems linked to a service.

As administrative data are collected as part of the day-to-day running of a service, they use fewer resources than special purpose surveys. If the government or other fund provider requires data collection as a condition of on-going funding to a service, the data collected by that service will tend to be more reliable, complete and of higher quality.

Information from administrative data collections is likely to be useful to a wide range of stakeholders such as service providers, higher-level bureaucrats, disability consumer groups, and researchers.

How can Administrative Data be used

Administrative data collections provide information on numbers and characteristics of service users, and the type, quantity and cost of services provided. Depending on the detail of the dataset and the complexity of the data collection format, a substantial amount of information about services, service users and service providers can be collected in this manner.

Administrative data of this sort can then be adopted by service agencies, planners, consumer advocacy groups, and funding department for a variety of purposes:

- To provide an evidence base to support planning for future service delivery (*e.g.*, by monitoring trends in client age or disability type);
- To indicate whether identified groups of people (*e.g.*, particular ethnic groups) are accessing services as much as expected;
- To monitor the cost-effectiveness of service provision;
- To support a budget submission for increased funding.

Moreover, with only minimal analysis, these data can answer basic administrative questions such as:

- How many people were supported by the service in a given time period, and what were their characteristics and support needs?
- What type of support was provided and received?
- What quantity of support-*e.g.*, in terms of staff hours-was provided and received?
- From whom was the support received (in terms of details of the service agency, such as size, staff profile, and hours)?
- What was the cost (total, per service type, per client) of providing these services?
- What were the outcomes for clients? (Examples of outcome questions include: Were clients satisfied with the services received? Were there increased levels of client participation in key life areas such as employment?)

In Australia, the Commonwealth/State/Territory Disability Agreement (CSTDA) funds a national program of disability support services for people with on-going support needs. Under the agreement, the Commonwealth Government is responsible for the overall planning and management of employment services, while the States and Territories are responsible for all other disability support services (including accommodation support, community access, community support, and respite services). All three jurisdictions share responsibility for advocacy, information, and print disability services. Services are provided by government and non-government service provision agencies.

'Persons with disabilities' are defined as people with a disability attributable to an intellectual, psychiatric, sensory, physical or neurological impairment or acquired brain injury (or some combination of these), which is likely to be permanent and to result in substantially reduced capacity in self-care, mobility or communication, requiring on-going or episodic support.

National Data on Disability Services

From 1995-2002, the National Minimum Data Set (NMDS) collected data about services received on a single 'snapshot' day each year. Since then, data have been collected on a full-year, on-going basis, so that the collection now includes data on all persons with disabilities who receive a CSTDA-funded service during the year. The NMDS collects the minimal, essential set of data about disability services and clients. It is a set of nationally-agreed upon data items, and an agreed method of collection and transmission.

The data collected includes:

- Demographic information, *e.g.*, age, sex, indigenous status
- Support needs and whether the person has a carer
- Primary disability and other disabilities
- Living arrangements and communication method
- Work, income, and funding, *e.g.*, main source income
- Services the person receives

It also contains data items on service provider agencies, including the type of service provided, staff hours, operating hours per day, and number of service users.

How the data is used: The CSTDA-NMDS data provides valuable information about those who are receiving services, their characteristics (age, disability type, and support needs) as well as trends in types of services used. Data for nine years is now available.

The data is used in funding negotiations between Commonwealth and State governments, and between departments and service providers. It is also used for developing national performance indicators, by which service outcomes can be monitored. The Australian Institute of Health and Welfare publishes a national data report on the collection each year.

JOINT USE OF CENSUS AND SAMPLE SURVEY-CENSUS SCREEN

use as sample frames for social surveys and instead use area-based samples in their household data collections to select respondents for their surveys. Persons with disabilities are a relatively small population and so obtaining a sufficient sample can be very costly.

Some countries use a small number of disability screening questions to provide the sampling frame for a follow-up survey. When this is done, it is important that the screening questions are effective in identifying as many persons with disabilities as possible. In particular, screening questions should attempt to minimize the number of "false negative" responses.

Another method to find a targeted sample inexpensively is to use the census as a screening device to identify a population of persons with disabilities who are then the sample for a follow-up (or "post-census") survey. As mentioned, relying on one or two screening questions will increase the number of false negatives. To compensate for this, it is wise to include in the follow-up survey a sufficient sample of negative responders to avoid biasing the survey population. Studies have shown that children with disabilities and people with mild disabilities are the most likely to be under-reported by the census screening questions.

Whenever the census is used as part of a data collection strategy, timing becomes a factor. In order to add questions into a census, steps must be taken several years in advance of the actual data collection. Furthermore, it will take one or two additional years to have the census data processed and available for post-census surveys or other uses. The different methods used by some countries in collecting disability data. Four of the nine countries (Cambodia, Fiji Islands, India, and Thailand) employed at least two methods of gathering disability data. Most countries in the Asia-Pacific region do not have registers of the population

Use of the Post-Census Survey Approach to Collect Disability Data in Canada. The post-census survey approach was used by Statistics Canada in its comprehensive surveys on disability following the 1986, 1991 and 2001 censuses.

The censuses included two screening questions on disability and impairment. In 2001, the questions were:

- Does this person have any difficulty hearing, seeing, communicating, walking, climbing stairs, bending, learning or doing any similar activities?
  Yes, sometimes
  Yes, often
  No
- Does a physical condition or mental condition or health problem reduce the amount or the kind of activity this person can do
  At home?
  At work or at school?
  In other activities, for example, transportation?

The purpose of these questions was not to estimate the prevalence of impairment or activity limitations. They merely defined a sample of individuals likely to have a disability. Following the census, a national sample survey based on census results was prepared. Those who screened positive in the census by answering *yes* to at least one of the questions were then asked detailed questions to confirm whether they had disability. A small sample of people who screened negative in the census was also included.

The post-census survey results provided far more accurate information on the prevalence of disability than possible with a census. The survey also provided detailed data about the nature of the disability and other characteristics of persons identified.

RELATABLE DATA

Table above Illustrates that each data collection method has its strengths and weaknesses concerning disability statistics. If more than one method is used, in a coordinated fashion, then the strengths of each method will be preserved and the weaknesses of each avoided. Deficiencies with census data, for example, can be addressed through disability surveys or disability modules in other surveys. The inadequacies with survey data can be partly overcome by using censuses, registered populations, and other administrative data.

Undoubtedly, a major challenge when assessing the life situation of persons with disabilities is locating and integrating data from various sources. First of all, there must be a variety of data sources to draw upon, and in many developing countries this is not true. Secondly, even if there are various data sources, the data must be relatable, that is based on a common conceptual framework and using comparable concepts and language. The ICF provides just what is required for relatability: a universal framework and a common language of disability.

Example of the use of data from a variety of sources to examine the unmet needs for disability services, showing in particular how survey and administrative data can be analysed together to provide information regarding the lives of persons with disabilities.

In Institute of Health and Welfare was commissioned to estimate levels of unmet need for disability services among persons with disabilities in Australia. Specifically, the project aimed to:

- Assess the effectiveness of previously allocated funding in reducing unmet need for disability services, by quantifying and describing additional services provided as a result of the funding; and
- Identify any remaining unmet need for disability accommodation, in-home support, day programs, respite services, and disability employment.

The two sources of data used were the NMDS data on services and consumers (the source of information on the supply of services); and disability population survey data (a source of information on unmet need for services).In conducting the analysis, the ICF was used as a common framework to which concepts and data items from these two data sources were mapped. Overlapping ICF concepts could then be identified in both data sources:

Disability services data collection	*Disability survey*
Impairments	
(*Need for assistance with .*) **Activities**: Self-care, Mobility Communication, other activities	(*Need for assistance with .*) **Activities**: Self-care, Mobility Communication, other activities
Participation	

Thus, a single 'indicator' of disability-"need for assistance"-was linked to the core purpose of a range of disability services. Estimation of the number of people in the Australian population with unmet needs for disability services was then based on an analysis of the survey data.

The analysis involved:

- Using survey data to identify people who reported a need for help with self-care, mobility, or communication. This group corresponds to the 'target population' for funded disability services; and
- Applying 'filters' to refine the estimate, focusing on people who were living in households (not in institutional accommodation) and who reported unmet need for formal help in addition to a need for frequent assistance.

The analysis produced the following estimates of unmet need across Australia:

- 12,500 people needing accommodation and respite services
- 8,200 places needed for community access
- 5,400 people needing employment support

Though thought to be too low, the findings informed negotiations between State, Territory and Commonwealth governments regarding funding for disability supports. This powerful analysis was only possible because common concepts were used defining the target population for disability services and the disability population survey. This was accomplished because both sets of concepts could be mapped to the ICF framework.

PREPARATION OF QUESTIONNAIRE AND SCHEDULE

QUESTIONNAIRE DESIGN

Questionnaires are an inexpensive way to gather data from a potentially large number of respondents. Often they are the only feasible way to reach a number of reviewers large enough to allow statistically analysis of the results. A well-designed questionnaire that is used effectively can gather information on both the overall performance of the test system as well as information on specific components of the system. If the questionnaire includes demographic questions on the participants, they can be used to correlate performance and satisfaction with the test system among different groups of users.

It is important to remember that a questionnaire should be viewed as a multi-stage process beginning with definition of the aspects to be examined and ending with interpretation of the results. Every step needs to be designed carefully because the final results are only as good as the weakest link in the questionnaire process. Although questionnaires may be cheap to administer compared to other data collection methods, they are every bit as expensive in terms of design time and interpretation.

The steps required to design and administer a questionnaire include:

- Defining the Objectives of the survey
- Determining the Sampling Group
- Writing the Questionnaire
- Administering the Questionnaire
- Interpretation of the Results

This document will concentrate on how to formulate objectives and write the questionnaire. Before these steps are examined in detail, it is good to consider what questionnaires are good at measuring and when it is appropriate to use questionnaires.

What can Questionnaires Measure

Questionnaires are quite flexible in what they can measure, however they are not equally suited to measuring all types of data. We can classify data in two ways,Subjective vs. Objective and Quantitative vs. Qualitative.

When a questionnaire is administered, the researchers control over the environment will be somewhat limited. This is why questionnaires are inexpensive to administer. This loss of control means the validity of the results are more reliant on the honesty of the respondent. Consequently, it is more difficult to claim complete objectivity with questionnaire data then with results of a tightly controlled lab test. For example, if a group of participants are asked on a questionnaire how long it took them to learn a particular function on a piece of software, it is likely that they will be biased towards themselves and answer, on average, with a lower than actual time. A more objective usability test of the same function with a similar group of participants may return a significantly higher learning time.

More elaborate questionnaire design or administration may provide slightly better objective data, but the cost of such a questionnaire can be much higher and offset their economic advantage. In general, questionnaires are better suited to gathering reliable subjective measures, such as user satisfaction, of the system or interface in question.

Questions may be designed to gather either qualitative or quantitative data. By their very nature, quantitative questions are more exact then qualitative. For example, the word "easy" and "difficult" can mean radically different things to different people. Any question must be carefully crafted, but in particular questions that assess a qualitative measure must be phrased to avoid ambiguity. Qualitative questions may also require more thought on the part of the participant and may cause them to become bored with the questionnaire sooner. In general, we can say that questionnaires can measure both qualitative and quantitative data well, but that qualitative questions require more care in design, administration, and interpretation.

When to use a Questionnaire

There is no all encompassing rule for when to use a questionnaire. The choice will be made based on a variety of factors including the type of information to be gathered and the available resources for the experiment. A questionnaire should be considered in the following circumstances.

- When resources and money are limited. A Questionnaire can be quite inexpensive to administer. Although preparation may be costly, any data collection scheme will have similar preparation expenses. The administration cost per person of a questionnaire can be as low as postage and a few photocopies. Time is also an important resource that questionnaires can maximize. If a questionnaire is self-administering, such as a e-mail questionnaire, potentially several thousand people could respond in a few days. It would be impossible to get a similar number of usability tests completed in the same short time.
- When it is necessary to protect the privacy of the participants. Questionnaires are easy to administer confidentially. Often confidentiality is the necessary to ensure participants will respond honestly if at all. Examples of such cases would include studies that need to ask embarrassing questions about private or personal behaviour.
- When corroborating other findings. In studies that have resources to pursue other data collection strategies, questionnaires can be a useful confirmation tools. More costly schemes may turn up interesting trends, but occasionally there will not be resources to run these other tests on large enough participant groups to make the results statistically significant. A

follow-up large scale questionnaire may be necessary to corroborate these earlier results.

DEFINING THE OBJECTIVES OF THE SURVEY

The importance of well-defined objectives can not be over emphasized. A questionnaire that is written without a clear goal and purpose is inevitably going to overlook important issues and waste participants' time by asking useless questions. The questionnaire may lack a logical flow and thereby cause the participant to lose interest. Consequential, what useful data you may have collected could be further compromised. The problems of a poorly defined questionnaire do not end here, but continue on to the analysis stage. It is difficult to imagine identifying a problem and its cause, let alone its solution, from responses to broad and generalizing questions. In other words, how would it be possible to reach insightful conclusions if one didn't actually know what they had been looking for or planning to observe.

A objective such as "to identify points of user dissatisfaction with the interface and how these negatively affect the software's performance" may sound clear and to the point, but it is not. The questionnaire designer must clarify what is meant by user dissatisfaction. Is this dissatisfaction with the learning of the software, the power of the software, of the ease of learning the software? Is it important for the users to learn the software quickly if they learn it well? What is meant by the software's performance? How accurate must the measurements be? All of these issues must be narrowed and focused before a single question is formulated. A good rule of thumb is that if you are finding it difficult to write the questions, then you haven't spent enough time defining the objectives of the questionnaire. Go back and do this step again. The questions should follow quite naturally from the objectives.

WRITING THE QUESTIONNAIRE

At this point, we assume that we have already decided what kind of data we are to measure, formulated the objectives of the investigation, and decided on a participant group. Now we must compose our questions.

If the preceding steps have been faithfully executed, most of the questions will be on obvious topics. Most questionnaires, however, also gather demographic data on the participants. This is used to correlate response sets between different groups of people. It is important to see whether responses are consistent across groups. For example, if one group of participants is noticeably less satisfied with the test interface, it is likely that the interface was designed without fair consideration of this group's specific needs. This may signify the need for fundamental redesign of the interface. In addition, certain questions simply may only be applicable to certain kinds of users. For example, if one is asking the participants whether they find the new tutorial helpful, we do not want to include in our final tally the responses of experienced users who learned the system with an older tutorial. There is no accurate way to filter out these responses without simply asking the users when they learned the interface.

Typically, demographic data is collected at the beginning of the questionnaire, but such questions could be located anywhere or even scattered throughout the questionnaire. One obvious argument in favour of the beginning of the questionnaire is that normally background questions are easier to answer and can ease the respondent into the questionnaire. One does not want to put off the participant by jumping in to the most difficult questions. We are all familiar with such kinds of questions.

It is important to ask only those background questions that are necessary. Do not ask income of the respondent unless there is at least some rational for suspecting a variance across income levels. There is often only a fine line between background and personal information. You do not want to cross over in to the personal realm unless absolutely necessary. If you need to solicit personal information, phrase your questions as unobtrusively as possible to avoid ruffling your participants and causing them to answer less than truthfully.

What kind of questions do we ask

In general, there are two types of questions one will ask, open format or closed format. Open format questions are those that ask for unprompted opinions. In other words, there are no predetermined set of responses, and the participant is free to answer however he chooses. Open format questions are good for soliciting subjective data or when the range of responses is not tightly defined. An obvious advantage is that the variety of responses should be wider and more truly reflect the opinions of the respondents. This increases the likelihood of you receiving unexpected and insightful suggestions, for it is impossible to predict the full range of opinion. It is common for a questionnaire to end with and open format question asking the respondent for her unabashed ideas for changes or improvements.

Open format questions have several disadvantages. First, their very nature requires them to be read individually. There is no way to automatically tabulate or perform statistical analysis on them. This is obviously more costly in both time and money, and may not be practical for lower budget or time sensitive evaluations. They are also open to the influence of the reader, for no two people will interpret an answer in precisely the same way. This conflict can be eliminated by using a single reader, but a large number of responses can make this impossible. Finally, open format questions require more thought and time on the part of the respondent. Whenever more is asked of the respondent, the chance of tiring or boring the respondent increases. Closed format questions usually take the form of a multiple-choice question. They are easy for the respondent, give

There is no clear consensus on the number of options that should be given in an closed format question. Obviously, there needs to be sufficient choices to fully cover the range of answers but not so many that the distinction between them becomes blurred. Usually this translates into five to ten possible answers per questions. For questions that measure a single variable or opinion, such as ease of use or liability, over a complete range (easy to difficult, like to dislike), conventional wisdom says that there should be an odd number of alternatives. This allows a neutral or no opinion response. Other schools of thought contend that an even number of choices is best because it forces the respondent to get off the fence. This may induce the some inaccuracies for often the respondent may actually have no opinion. However, it is equally arguable that the neutral answer is over utilized, especially by bored questionnaire takers. For larger questionnaires that test opinions on a very large number of items, such as a music test, it may be best to use an even number of choices to prevent large numbers of no-thought neutral answers.

Closed format questions offer many advantages in time and money. By restricting the answer set, it is easy to calculate percentages and other hard statistical data over the whole group or over any subgroup of participants. Modern scanners and computers make it possible to administer, tabulate, and perform preliminary analysis in a matter of days. Closed format questions also make it easier to track opinion over time by administering the same questionnaire to different but similar participant groups at regular intervals. Finally closed format questions allow the researcher to filter out useless or extreme answers that might occur in an open format question.

Whether your questions are open or closed format, there are several points that must by considered when writing and interpreting questionnaires:

Clarity

This is probably the area that causes the greatest source of mistakes in questionnaires. Questions must be clear, succinct, and unambiguous. The goal is to eliminate the chance that the question will mean different things to different people. If the designers fails to do this, then essentially participants will be answering different questions.

To this end, it is best to phrase your questions empirically if possible and to avoid the use of necessary adjectives. For example, it asking a question about frequency, rather than supplying choices that are open to interpretation such as:

- Very Often
- Often
- Sometimes
- Rarely
- Never

It is better to quantify the choices, such as:

- Every Day or More
- 2-6 Times a Week
- About Once a Week
- About Once a Month
- Never

There are other more subtle aspects to consider such as language and culture. Avoid the use of colloquial or ethnic expressions that might not be equally used by all participants. Technical terms that assume a certain background should also be avoided.

Leading Questions

A leading question is one that forces or implies a certain type of answer. It is easy to make this mistake not in the question, but in the choice of answers. A closed format question must supply answers that not only cover the whole range of responses, but that are also equally distributed throughout the range. All answers should be equally likely.

An obvious, nearly comical, example would be a question that supplied these answer choices:

- Superb
- Excellent
- Great
- Good
- Fair
- Not so Great

A less blatant example would be a Yes/No question that asked:

- Is this the best CAD interface you have every used?

In this case, even if the participant loved the interface, but had an favourite that was preferred, she would be forced to answer No. Clearly, the negative response covers too wide a range of opinions.

A better way would be to ask the same question but supply the following choices:

- Totally Agree
- Partially Agree
- Neither Agree or Disagree
- Partially Disagree
- Totally Agree

This example is also poor in the way it asks the question. It's choice of words makes it a leading question and a good example for the next section on phrasing.

Phrasing

Most adjectives, verbs, and nouns in English have either a positive or negative connotation. Two words may have equivalent meaning, yet one may be a compliment and the other an insult. Consider the two words "child-like" and "childish", which have virtually identical meaning. Child-like is an affectionate term that can be applied to both men and women, and young and old, yet no one wishes to be thought of as childish.

In the above example of "Is this the best CAD interface you have every used?" clearly "best" has strong overtones that deny the participant an objective environment to consider the interface. The signal sent the reader is that the designers surely think it is the best interface, and so should everyone else. Though this may seem like an extreme example, this kind of superlative question is common practice.

A more subtle, but no less troublesome, example can be made with verbs that have neither strong negative or positive overtones.

Consider the following two questions:

- Do you agree with the Governor's plan to oppose increased development of wetlands?
- Do you agree with the Governor's plan to support curtailed development of wetlands?

They both ask the same thing, but will likely produce different data. One asks in a positive way, and the other in a negative. It is impossible to predict how the outcomes will vary, so one method to counter this is to be aware of different ways to word questions and provide a mix in your questionnaire. If the participant pool is very large, several versions may be prepared and distributed to cancel out these effects.

Embarrassing Questions

Embarrassing questions dealing with personal or private matters should be avoided. Your data is only as good as the trust and care that your respondents give you. If you make them feel uncomfortable, you will lose their trust. Do not ask embarrassing questions.

Hypothetical Questions

Hypothetical are based, at best, on conjecture and, at worst, on fantasy.

I simple question such as:

- If you were governor, what would you do to stop crime?

This forces the respondent to give thought to something he may have never considered. This does not produce clear and consistent data representing real opinion. Do not ask hypothetical questions.

Prestige Bias

Prestige bias is the tendency for respondents to answer in a way that make them feel better. People may not lie directly, but may try to put a better light on themselves. For example, it is not uncommon for people to respond to a political opinion poll by saying they support Samaritan social programs, such as food stamps, but then go on to vote for candidates who oppose those very programs. Data from other questions, such as those that ask how long it takes to learn an interface, must be viewed with a little skepticism. People tend to say they are faster learners than they are.

There is little that can be done to prevent prestige bias. Sometimes there just is no way to phrase a question so that all the answers are noble. The best means to deal with prestige bias is to make the questionnaire as private as possible. Telephone interviews are better than person-to-person interviews, and written questionnaires mailed to participants are even better still. The farther away the critical eye of the researcher is, the more honest the answers.

NOW WHAT

Now that you've completed you questionnaire, you are still not ready to send it out. Just like any manufactured product, your questionnaire needs to go through quality testing. The major hurdle in questionnaire design is making it clear and understandable to all. Though you have taken great care to be clear and concise, it is still unreasonable to think that any one person can anticipate all the potential problems. Just as a usability test observes a test user with the actual interface, you must observe a few test questionnaire takers. You will then review the questionnaire with the test takers and discuss all points that were in any way confusing and work together to solve the problems. You will then produce a new questionnaire. It is possible that this step may need to be repeated more than once depending on resources and the need for accuracy.

CONCLUSIONS

Questionnaire design is a long process that demands careful attention. A questionnaire is a powerful evaluation tool and should not be taken lightly. Design begins with an understanding of the capabilities of a questionnaire and how they can help your research. If it is determined that a questionnaire is to be used, the greatest care goes into the planning of the objectives. Questionnaires are like any scientific experiment. One does not collect data and then see if they found something interesting. One forms a hypothesis and an experiment that will help prove or disprove the hypothesis.

Questionnaires are versatile, allowing the collection of both subjective and objective data through the use of open or closed format questions. Modern computers have only made the task of collecting and extracting valuable material more efficient. However, a questionnaire is only as good as the questions it contains. There are many guidelines that must be met before you questionnaire can be considered a sound research tool.

The majority deal with making the questionnaire understandable and free of bias. Mindful review and testing is necessary to weed out minor mistakes that can cause great changes in meaning and interpretation. When these guidelines are followed, the questionnaire becomes a powerful and economic evaluation tool.

PRECAUTIONS

DATA PREPARATION

Data Preparation involves checking or logging the data in; checking the data for accuracy; entering the data into the computer; transforming the data; and developing and documenting a database structure that integrates the various measures.

Logging the Data

In any research project you may have data coming from a number of different sources at different times:

- Mail surveys returns
- Coded interview data
- Pretest or posttest data
- Observational data

In all but the simplest of studies, you need to set up a procedure for logging the information and keeping track of it until you are ready to do a comprehensive data analysis. Different researchers differ in how they prefer to keep track of incoming data.

In most cases, you will want to set up a database that enables you to assess at any time what data is already in and what is still outstanding.

The data analyst should always be able to trace a result from a data analysis back to the original forms on which the data was collected. A database for logging incoming data is a critical component in good research record-keeping.

Checking the Data For Accuracy

As soon as data is received you should screen it for accuracy. In some circumstances doing this right away will allow you to go back to the sample to clarify any problems or errors.

There are several questions you should ask as part of this initial data screening:

- Are the responses legible/readable?
- Are all important questions answered?
- Are the responses complete?
- Is all relevant contextual information included (*e.g.,* data, time, place, researcher)?

In most social research, quality of measurement is a major issue. Assuring that the data collection process does not contribute inaccuracies will help assure the overall quality of subsequent analyses.

Developing a Database Structure

The database structure is the manner in which you intend to store the data for the study so that it can be accessed in subsequent data analyses. You might use the same structure you used for logging in the data or, in large complex studies, you might have one structure for logging data and another for storing it. As mentioned above, there are generally two options for storing data on computer -- database programs and statistical programs. Usually database programs are the more complex of the two to learn and operate, but they allow the analyst greater flexibility in manipulating the data.

In every research project, you should generate a printed codebook that describes the data and indicates where and how it can be accessed.

Minimally the codebook should include the following items for each variable:

- Variable name
- Variable description

- Variable format (number, data, text)
- Instrument/method of collection
- Date collected
- Respondent or group
- Variable location (in database)
- Notes

The codebook is an indispensable tool for the analysis team. Together with the database, it should provide comprehensive documentation that enables other researchers who might subsequently want to analyse the data to do so without any additional information.

Entering the Data into the Computer

There are a wide variety of ways to enter the data into the computer for analysis. Probably the easiest is to just type the data in directly. In order to assure a high level of data accuracy, the analyst should use a procedure called double entry. In this procedure you enter the data once. Then, you use a special program that allows you to enter the data a second time and checks each second entry against the first. If there is a discrepancy, the program notifies the user and allows the user to determine the correct entry. This double entry procedure significantly reduces entry errors. However, these double entry programs are not widely available and require some training.

An alternative is to enter the data once and set up a procedure for checking the data for accuracy. For instance, you might spot check records on a random basis. Once the data have been entered, you will use various programs to summarize the data that allow you to check that all the data are within acceptable limits and boundaries. For instance, such summaries will enable you to easily spot whether there are persons whose age is 601 or who have a 7 entered where you expect a 1-to-5 response.

Data Transformations

Once the data have been entered it is almost always necessary to transform the raw data into variables that are usable in the analyses. There are a wide variety of transformations that you might perform.

Missing Values

Many analysis programs automatically treat blank values as missing. In others, you need to designate specific values to represent missing values. For instance, you might use a value of -99 to indicate that the item is missing. You need to check the specific program you are using to determine how to handle missing values.

Item Reversals

On scales and surveys, we sometimes use reversal items to help reduce the possibility of a response set. When you analyse the data, you want all scores for scale items to be in the same direction where high scores mean the same thing and low scores mean the same thing. In these cases, you have to reverse the ratings for some of the scale items. For instance, let's say you had a five point response scale for a self esteem measure where 1 meant strongly disagree and 5 meant strongly agree.

One item is "I generally feel good about myself." If the respondent strongly agrees with this item they will put a 5 and this value would be indicative of higher self esteem. Alternatively, consider an item like "Sometimes I feel like I'm not worth much as a person."

Here, if a respondent strongly agrees by rating this a 5 it would indicate low self esteem. To compare these two items, we would reverse the scores of one of them (probably we'd reverse the latter item so that high values will always indicate higher self esteem). We want a transformation where if the original value was 1 it's changed to 5, 2 is changed to 4, 3 remains the same, 4 is changed to 2 and 5 is changed to 1. While you could program these changes as separate statements in most program, it's easier to do this with a simple formula like:

New Value = (High Value + 1)-Original Value

In our example, the High Value for the scale is 5, so to

get the new (transformed) scale value, we simply subtract each Original Value from 6 (*i.e.*, 5 + 1).

Scale Totals

Once you've transformed any individual scale items you will often want to add or average across individual items to get a total score for the scale.

Categories

For many variables you will want to collapse them into categories. For instance, you may want to collapse income estimates (in dollar amounts) into income ranges.

TIPS FOR DATA COLLECTION

As we have worked with school districts to implement teaching and technology surveys, here are a few tips we've developed to create teacher surveys that work. Whether you create your own paper-based survey or use an on-line tool, keeping these basic points in mind will help you get the largest return for your effort.

Challenge your Assumptions

It is tempting to make assumptions about why things are the way they are—for instance, why some teachers are excited about educational technology and others aren't. Issues of training access, skills, and attitudes are complex and usually there is more to know than meets the eye. Collect data from your teachers, students, administrators, and community, and then use that data to drive technology planning and implementation efforts.

Collect Data from a Wide Audience

Remember that you have a lot of stakeholders to your technology implementation work; therefore, you need to collect data from all of these stakeholder populations. Just as important as sample size is sample scope — think about all of the individuals who will be affected by the issue at hand, such as a new technology plan.

Collect Data that can't be Directly Observed

The *best* data is often that which provides insights into attitudes and opinions. You tend to get this data as notes from discussions/interviews or written comments to open-ended survey questions.

Sure, this is harder to tabulate, but it provides rich insight into what people are *thinking* versus just what they are doing (or saying that they are doing). Besides, you can get the "doing" data without actually surveying... what you want to collect is that information you do not otherwise see. Data is more than quantitative information and "yes/no" answers to survey questions.

Always have more than one Data Source

Surveys alone cannot provide all of your relevant data. Therefore, always have more than one data source (*e.g.*, surveys AND interviews AND observations) and then create your analysis from a comparison between the sources. Self-reported surveys are always subjective to some degree, so having lots of other data sources for comparison will help to paint the clearest possible picture of the people behind the survey responses.

Have Multiple Survey-writers

When creating surveys, try to have more than one person involved in writing questions. Have people read each other's work.

This helps address the fact that most questions can be perceived of in more than one way. You need to test your survey language among the survey-writers before springing this on your survey population.

Set Reasonable Expectations

You should try to "encourage" the reluctant to participate (as they add a very interesting and worth while aspect to your overall data picture), but you really cannot force people to participate. In most cases (*i.e.*, when you have more than 100 teachers), 50% is fine and even 25% is respectable. Also, remember that quantity is only one aspect of good data...scope (*i.e.,* including all stakeholder communities) is just as important.

Mailbox Surveys do not Work

Teachers will not respond in large numbers to surveys that are simply left in their mailboxes. When your survey becomes just one more piece of junkmail, it will most often be ignored. One way to increase your returns is to let your teachers complete your survey electronically, through a web-based form.

We've seen districts come away with higher returns when surveys are administered via the web. If you can take a few extra minutes at a faculty meeting, have everyone log on at the same time and answer the survey questions right then and there.

Choose your Survey Questions Carefully

You probably will have no more than one chance per year to hear from your stakeholders through a survey. Make it count! Be as specific as possible with your survey questions and make sure you get to the heart of the issues that are important to your team. Consult other school districts' surveys for ideas, but write questions that make sense for your school district.

Be Prepared to Report back Results and take Action

People will always be more willing to participate in a survey (and participate again later) if they know that the results will be communicated to them and that the data they provide will make a difference in future plans.

Survey, and then Survey Again

Surveys often play a role at different stages throughout an evaluation process. For example, you may take a survey to collect baseline data that informs a planning process, then take another survey to measure how things are going the following year.

4

Nano Research Analysis

ANALYSING DATA

Langley further notes that "In theorizing from process data, we should not have to be shy about mobilizing both inductive (data-driven) approaches and deductive (theory-driven) approaches iteratively or simultaneously as inspiration guides us". The key anchor points depicted in the table are according to Langley to fix attention on some point(s) that help to structure the material, but also to determine which elements will receive less attention. Interestingly the different approaches produce different levels of accuracy, generality and simplicity. The data analysis process carried out in this study exploiting the above-depicted Langley's strategies is subsequently described:

All the cases were analysed several times in order to minimize the possibility of the understanding changing during the reading and analysis process. When using transcriptions the categorization process took five weeks, but the tentative computerized analysis work used somewhat less time.

During the *open coding* phase which was first done assisted by software several tentative categories were established beforehand and during the analysis process itself new (sub) categories were added.

Among these categories were *e.g.* business processes, software processes, legal processes and contracts but as this approach was not fruitful it was abandoned, ref. below *Alternate templates strategy*. Next the approach was totally turned around and the analysis work started by defining all process activities from the data, and there was a change to use the axial coding method in this phase. In axial coding the researcher selects the main characteristics as the object of the analysis and the coding is done using these axles. Strauss and Corbin define the axial coding to be "the act of relating categories to subcategories along the lines of their properties and dimensions".

The activities thus form the smallest unit of analysis and these were subsequently classified with four attributes:

- *Phases* comprised of negotiation, definition, production and assessment classes.
- *Interaction process* comprised of exchange, adaptation and coordination classes.
- *Ford's stage* comprised of pre-relationship, exploratory, developing, and stable classes.
- *Motor* comprised of prescribed and constructive classes.

Further to these elements the following attributes to help to scrutinize the processes were attached: inputs, outputs, owner and events.

All the process activities were marked and translated into English and subsequently they were inserted case by case into individual Excel tables in order to help the analysis work by enabling the data to be viewed in more compact form, Appendix D. After this the different classes were inserted together in one Excel table to make it possible to order the activities in different orders to see any interrelationships and patterns, Appendix E. These shorthand tables were later modified to become countable in order to be able to derive some coarse statistics out of the data.

- *Narrative strategy* was used translating short English synopses from each transcribed case material. The synopses included contractual issues and contract process activity descriptions that later formed the base for the detailed and numbered process activity lists, Appendix D. In one case a short written narrative and descriptive story was written in order to highlight the interesting situation where the focal company had a completely immature contracting process, Appendix H.

- *Quantification strategy* was used to some extent when the process activity lists were transformed into calculable forms, Appendix D and E. Tabularisations were done using the Excel software and the worksheets were then capitalized to produce simple 3x3 matrixes from different interesting phenomena. In this context it was not the case of using formal and extensive statistical methods, as the scope of the available data did not permit this.
- *Alternate templates strategy* were used during preliminary tentative computer assisted analysis work. The data categorization was started by reading the transcriptions through several times. First a robust tree structured pre-categorization was created, and during the analysis process augmented up until 15 main categories at the root level and approximately 65 subcategories at the so-called child level after becoming acquainted more with the material. However, soon this seemed to be too hard to use, as well as being too atomized. There was an evident risk of losing the overview of the whole data. So the next step was to drop the number of the main categories to 10 and the subcategories to 51. This pre-categorization was also found to be too detailed. The second classification approach was to classify all incidents and expressions on different contracts, but this perspective would have been useful if the viewpoint of the research would have only been from the legal and contractual perspective. It did not help the dynamic processual study. Reading and pondering about the empirical data with different tools and from different angles gave more and more insight into it. However, two problems could be expressed and they are, first, that the computerized analysis easily produces too many categories and,

secondly, technically the computer screen is not so agreeable and handy to use in the process as normal A4-paper. Although the time spent with the empirical material and the software was not in any way wasted time. After this experiment the researcher started to read the transcription papers in order to categorize the process elements activities. The analysis made reading the papers was also more illustrative and it was also easier to keep track of the whole – maintaining a helicopter perspective – when analyzing and classifying the smaller process activities.

- *Grounded theory strategy* was used extensively as can be This was the main method used in this study. It was also best suited for this type of research where the researcher approached new and untouched fields. The objective of the grounded theory based research approach is to generate theory rather than to test it, this gives the researcher analytical tools for handling a large empirical data base, and helps him to find alternative meanings for the phenomena (Rantala 1999). This enables the work to be at the same time systematic as well as creative and during the whole research process the researcher strives to identify, develop and proportion the concepts, *i.e.* the building blocks of theory.
- *Visual mapping strategy* was used to some extent by describing every case as the example in Appendix I shows.
- *Temporal bracketing strategy* is used to visualize the contracting process as well as other contractual issues in the different business modes.
- *Synthetic strategy* is put to use in Chapter 7 to elaborate the Dynamic contracting process model. According to Langley the process theorizing method produces an enriched understanding and explanation, however, it often lacks predictive power (ibid). The synthetic method uses the process as a

whole (unit of analysis) and constructs global measures from the detailed event data in order to be able to describe and compare the corresponding processes from different cases.

The synthetic strategy requires:

- Clearly defined process boundaries – in this focal case it is the software contracting process,
- Abstraction level that is high enough to permit the comparison between several cases – in Chapter 4 elaborated conceptual model fills these requirements well, and
- The strategy requires a sufficient number of cases – this study used twelve different cases that gave enough data to infer rational conclusions about the contract process unfolding.

To use the Möller and Wilson model as a base in constructing the software contracting process model was quite a straightforward effort. In this study the customer's viewpoint was deliberately left out, as the main focus was on the software company's contracting process and its behaviour linked to software development processes. Therefore the customer context in the Möller and Wilson model was left for less scrutiny.

From this model the environmental context, supplier and task specific characteristics, organizational characteristics, interaction processes and outcomes elements were used. In the elaborated software business model these elements were specified as reflecting the business at issue.

This study tries to capitalize on the above-mentioned seven methods to a varying extent in order to cover and get the relevant issues from the empirical data from as many angles as possible and in as many ways as possible. This data "massaging" helped the researcher to get better acquainted with the data as thoroughly as possible. To sum up, from the above-discussed strategies depicted by Langley, the *grounded theory* was used vitally in this research. The other strategies were more or less applied to enhance the understanding of and familiarize with the focal data.

CODING

CODING AND ANALYZING THE DATA

The danger here, as Pettigrew refers to it, is "death by data asphyxiation." The information processing requirements of large, complex data sets quickly exceed the capabilities of even the most experienced researchers. Organizing and evaluating these data becomes a very challenging task; in fact, new strategies and techniques for coding and analyzing these type of data often have to be devised. Although we uncovered no studies in New Public Management that can serve as exemplars in this regard, we offer two strategies that might be helpful to researchers of New Public Management at this stage in its evolution.

The first, event coding and sequence analysis, is ideal for longitudinal case studies that investigate any type of change process, especially those that rely on real-time observations and permit the transformation of qualitative data into quantitative measurements. Network analysis, the second strategy, is particularly useful in exploring how interactions and relational patterns change over time. We highlight only the basic steps in these strategies below; since each will be addressed in greater detail in the following section when we describe two research programs in some depth.

Incident Coding and Sequence Analysis

This strategy has evolved from longitudinal studies that investigate the innovation process (Van de Ven, et.al., 1989; Van de Ven and Poole, 1990).

In the course of their studies, seven steps emerged for coding and analyzing the data:

- Define the qualitative datum as an incident, bracket raw data collected from the field into these incidents, and entèr this information into a qualitative incident data file.
- Evaluate the reliability and validity of classifying raw data into incidents by (a) achieving consensus and

consistent interpretations of decision rules among at least two researchers performing this task, and (b) asking organizational participants to review the chronological list of incidents that occurred in their innovation or change effort.

- Code each incident in terms of the presence or absence of theoretical events constructs, and add these codes to the incident data file.
- Evaluate the reliability and validity of the event coding scheme by following conventional procedures for establishing construct validity and interrater reliability of measures.
- Transform the qualitative codes into dichotomous variables, or a bit-map event sequence data file, which permits time-series analysis of process theories of organizational change or development.
- Analyse temporal relationships between variables in the event sequence data file using a variety of statistical time series procedures appropriate to the theoretical question at hand. Enrich the interpretation of statistical results by reading and content analyzing the relevant sequence of incidents in the qualitative data file (developed in steps 1 to 3 above).
- Analyse developmental patterns or phases in organizational change or innovation by defining and examining coherent patterns of activity among temporal events in the incident data file".

Coding, Database Construction and Network analysis

Network analysis is the research methodology par excellence for studies that focus on the organization of social relationships. In contrast to all other social science methodologies, where the measurement unit (and unit of analysis) is centered on or derived from the individual, in network analysis the measurement unit is relational— the social tie or bond between two actors i and j. The following procedures are central:

- To set up the sociometric data for structural analysis, a dyad file must be created. The basic record in a dyad file contains data on the relation between a given pair of individuals. For each dyadic relations there are two records: one record containing the data on i's perception of the relation with j, and a second record containing j's perception of its relation to i. For example, in a group of ten individuals, the data file will contain ninety dyadic records or (N(N-1)) relations.
- For longitudinal studies, each time period over which the relations are mapped must be signified and separated in the database by a wave identifier.
- For structural analysis the raw relational data must be converted, by recoding into binary code (1 for the presence of a relation or tie; 0 for its absence or otherwise), into an adjacency matrix showing all possible pair-wise relations for a given relational content.
- For each different content analysed (*e.g.* the bond could represent power, affect etc.) or for different values of the relations of a given content, a new adjacency matrix must be constructed.
- Depending upon the kind of structural organization expected or predicted by the substantive theory, the sociomatrix is then subjected to a structural analysis using one or more structural analysis programs. These programs analyse the relations in the sociomatrix and provide structural measure of various patterns or arrangements of the relations, either for the whole social unit or for various subsets of individuals or relational entities.

DATA CODING AND ANALYSIS

Extensive techniques have been developed to codify procedures for handling longitudinal panels of quantitative data, including constructing computer data files and analyzing longitudinal data (Tuma and Hannan, 1984; Van de Ven and Chu, 1989). Less has been written on coding and analyzing qualitative data, especially those interested in identifying patterns of change. The following steps were devised in the course of the MIRP research project.

Defining a Datum as an Incident

An incident is defined as a recurrence or change in any one of the five core concepts (units) in the MIRP framework: innovation idea, people, transactions, context, and outcomes. When an incident is identified, it is described by a "bracketed string of words" which include: date of occurrence, the actor(s) or object(s) involved, the action or behaviour that occurred, the consequences (if any) of the action, and the source of the information. These "bracketed string of words about a discrete incident" are then entered into a qualitative incident data file.

Reliability and Validity of Transforming Raw Data Into Incidents

This step attempts to establish the reliability of classifying raw data into incidents. The process begins with at least two researchers who enter incidents from raw data sources into the data file. The two researchers have to have a consensus on the application of decision rules to the incidents. However, instead of just relying on researchers' classifications and agreement, practitioners also are asked to review the resultant list of incidents and indicate if any incidents are missing or incorrectly described.

This additional effort seeks to test empirically whether researchers' classifications and codes are consistent with practitioners' perception of events. Based on practitioners' feedback, revisions are made if they conform to the decision rules for defining each incident.

If evidence reveals an inconsistency between researchers' and practitioners' interpretations, then researchers can still make claims about the meaning of incidents from their theoretical perspective, but claims about practitioners' reality of the incident are not appropriate. It is acknowledged that the resultant list of incidents does not represent the population of occurrences in the development of an innovation, but instead represent a sample of incidents of what happened over time.

Coding Incidents into Event Constructs

A list of incidents is a qualitative indicator of what happens in the development of an innovation, but one additional step is needed. Researchers must code the incidents into theoretically-meaningful event constructs. MIRP researchers used the core concepts and developed multiple variables on which to code them into event constructs.

For example, when incidents provide evidence of results, they are coded as representing either a positive event construct (good news or successful accomplishment), negative event construct (bad news or instances of failure or mistakes), or mixed event construct (neutral or ambiguous news indicating elements of both success and failure).

Assessing Reliability and Validity of Coding Scheme

The actual coding of incidents into event constructs is performed independently by two or more researchers. This enables the researchers to compute inter-rater reliability.

Transforming Coded Incidents into Bit Maps for Time Series Analysis

The next step in the procedure is the transformation of coded incidents or event constructs into what the researchers call a "bit map." A bit map is a matrix of rows which represent the incidents listed in chronological order and rows which represent the variables representing all of the event constructs. Each event construct of an incident is coded into a dichotomous variable of 1 (change occurred) or 0 (no change occurred). This transformation of qualitative data into quantitative data permits the application of various statistical techniques to examine time-dependent patterns of relation among the event constructs.

Analysis of Temporal Relationships in Bit-Map Data

The stage is now set for examining temporal relationships and patterns among the variables in the development of innovation. The family of methods concerned with the problem of determining the temporal order among events is called sequence analysis (Abbott, 1984). It examines similarities and differences between discrete events. The bit map files can be analysed with a variety of statistical methods to identify time-dependent patterns among the dimensions coded as 1's and 0's.

The MIRP studies have utilized chi-square test and log-linear models to examine probabilistic relationships between categorical independent and dependent variables, Granger causality and vector autoregression to identify possible causal relationships between bit-map variables, and time series regression analysis on incidents aggregated into fixed temporal intervals to test specific process models. All of these methods attempt to detect bivariate relationships between coded event variables.

Analyzing Developmental Patterns or Phases in Temporal Data

This step in the analysis uses a multi-variate technique called phase analysis. Its purpose is to identify and compare developmental patterns or stages in the temporal sequence of data. This technique can be used to both develop and test models (hypotheses) about development. One particular advantage is that it can evaluate more than one process model. For example, the MIRP researchers used this technique to compare and contrast two models of the innovation process.

The phase analysis technique requires the researcher to conceptually define discrete phases of innovation activity and create a phase map. The next step is to analyse sequences and properties of the phases and to identify any meaningful patterns. MIRP researchers focused on two kinds of patterns — the types of sequences and the structural properties of sequences.

CODING AND ENTERING DATA

Coding is the development and use of a language that will be used to transfer data from the instument which was employed in the data collection process to a"codebook" or directly to the computer in a form that is appropriate for data analysis and reporting results. For the purpose of this course, this means making decisions about how to represent the data you collected on questionnaires in a way that the SPSS program will use effectively to do analysis. Since numbers usually work best as the way to represent the different responses made by the people you surveyed, the task is to find ways to code every response with a number.

Entering data is then done by putting the appropriate number for each datum (the singular of data) into an appropriate place in a data file. The first computer assignment requires you to build a data file which contains coded responses from at least fifty of the people who completed your questionnaire. There are some concepts which students need to be very clear about before they begin the coding process. These include the differences between variables and values.

Variables are "logical groupings of attributes." Values are the attributes thus grouped. So the variable gender has two attributes or values: female and male (the way it is usually constructed in our society). The variable age has many possible values: 18, 19, 20, 21, 34, 67, etc. It is important to remember that a variable needs more than one value. If all your respondents are social work majors, then major is not a variable. Don't waste your time coding and entering that data.

So each variable which you measured in your survey needs to be coded. If you start with an attitude, such as the first item on the demonstration questionnaire, "My heart goes out to people in wheelchairs," you need to have a code for each of the possible values: strongly agree, agree, undecided, disagree, and strongly disagree.

One way to do this, which I think is the easiest way, is to start with one and use as many numbers as you need, so the coding becomes:

1. For "strongly agree,"

2. For "agree,"
3. For "undecided,"
4. For "disagree," and
5. For "strongly disagree."

These are entirely arbitrary. You can reverse these or use any numbering scheme you want, but you need to be sure that you remember what the codes mean.

A good way to remember is to prepare a codesheet. This is best done by using a blank copy of the questionnaire which you used to collect data. It will give you a place to write variable names which are explained on the handout for the first computer assignment. Also, write down the numbers you use as codes for each value of every variable.

Variables that have the same values will be coded identically. Since all of the attitude items on the demonstration questionnaire have the same choices, they would all be coded the same:

1. For "strongly agree,"
2. For "agree,"
3. For "undecided,"
4. For "disagree," and
5. For "strongly disagree."

The next group of items, behaviour variables, are all Yes/No questions. These can be coded 1 for "Yes" and 2 for "No." It is OK for the 1 to be the code for "strongly agree" for one variable and for 1 to mean "Yes" for another. Later we will tell the computer what everything means.

It is usually the case that the values for each demographic will be coded differently, so that gender has the codes 1 for "Female" and 2 for "Male" (they can be in the other order if that's the way they are on your instrument); marital status would be: 1 for "Single," 2 for "Married," etc. A variable such as age, which is already represented by numbers, uses those numbers as the code, so, for example, the written in age of "22" is coded 22.

Other open-ended responses, such as major, are a little more complicated. Often we just code the first response we get as 1, the second as 2, and so on. However, you may want to reserve the low numbers for frequently occurring values. For example, if you have several social work majors in your sample, use 1 for "social work" since it is easy to remember. Open-ended items that have more individualistic responses, such as "What else do you think about this topic?" need to be coded more creatively. Remember: *write things down!* Keep a record of these codes on your codesheet so that you can remember what they mean.

Once you are clear about the coding process you are ready to begin data entry. The first computer assignment will help you with this. Be sure to ask in class if you have any questions as you begin that assignment.

EDITING AND TABULATION OF DATA

TABULATION OF DATA

In your proposal, you should discuss editing and tabulating data immediately after data collection procedures. Although qualitative methods are being increasingly used in operations research, most OR studies still involve quantitative analysis that requires statistical manipulation of the information collected. First, you need to convert the information into a form that will allow it to be analysed. Second, you must specify the statistical manipulations to be performed. Finally, you need to present the important findings resulting from these manipulations in a report or series of reports.

Preparing Tabulations

Any recently produced desktop computer probably has the hardware capability needed to process an operations research data set. However, unless the computer you use is located at a research organization or a health program evaluation unit, it may not have the software needed for statistical analysis. This is not a problem when data consist only of service statistics from a small number of service delivery points, or when modeling or conducting a cost analysis; in both cases, spreadsheets are adequate for analyzing OR data. However, it's more likely that you will need to perform many statistical tests, analyse survey data, or work with a very large data set and will need more powerful statistical software.

If, in response to question 110, the respondent states that he received his last HIV test at Central Hospital, the interviewer would circle the number 2. If the answer is the military camp, the interviewer would circle 4. Before beginning computerized data entry, you should check all questionnaires to make sure that the interviewers have recorded a response to each question.

If the number of categories for a particular variable (including, if relevant, codes for "nonresponse," "not applicable," "don't know," and "other") is less than 10, numerical codes should be single-digit numerals. If the total possible number of categories is between 10 and 99, two-digit codes should be used instead. For some variables, it may be necessary to use three-digit, four-digit, or even larger codes; for example, calendar dates typically require four or more digits.

DATA EDITING

Coded data need to be entered into the computer with a minimum of typing errors and then edited to correct any errors in the data. In entering data, the researcher should use the data verification procedures available with most statistical packages. In verification, the same data are entered twice. The verification program indicates discrepancies in the numbers entered. In the example above, the first data entry clerk might have entered the number 3. However, the second time the response is entered, it may be entered as 1. When such discrepancies occur, the program signals the data entry person to check the data entry form for the correct number. In addition to verification, the researcher should check for the following types of errors:

Ilegal" codes

Values that are not specified in the coding instructions.

For example, a code of "7" in question 110 above would be an illegal code. The best way to check for illegal codes is to have the computer produce a frequency distribution and check it for illegal codes. This transformed variable might be transformed even further for certain kinds of additional analysis. For example, if you want to cross-tabulate age by other variables, it is preferable to limit the age distribution to relatively few age categories (usually five- or ten-year categories) or even to dichotomize (for example, ages 15–29 and ages 30 or more). You can use several methods to transform variables, the most common of which are listed below.

- *Recodes:* In recoding, category labels are changed. This technique is used to "collapse" large numbers of variable categories into smaller numbers. For example, single years of age can be collapsed and transformed into age categories, such as ages 15– 19, 20–24, and 25–29.
- *Counts:* If you are collecting information on whether the respondents have ever used any of eight services for persons with HIV/AIDS, you might want to count the number of services ever used by each respondent. Thus, you could generate a new variable that might be called "Number of Services Ever Used."
- *Conditional Transformations*: When the nature of the transformation of one variable depends on the second variable, conditional transformations may be useful. *For instance, suppose you asked respondents three questions*:
  - Did you hear the partner reduction radio message in July?
  - How many casual sex partners did you have between April and June?
  - How many casual sex partners did you have between August and October?

CHARTS AND DIAGRAMS

INTRODUCTION

A chart is a technique of displaying data using pictures

and graphical representations instead of numbers or simple words.

It works by drawing figures that would represent numbers, adding colours and shapes to the information presented.

Good created and formatted charts can help people and businesses make decisions based on the impact that the images provide. Data analysis on charts is done using graphics that present pictures.

In addition to the pictures, you can add words, also called labels to indicate what the pictures represent. Because a chart is used to present data in a graphical format, before creating a chart, you should plan it.

That is, you should prepare it. There are two pieces of information you should have before starting: The numbers that you want to represent and the type of chart you want to use.

CREATING A CHART

The information used to create a chart usually come from two or more cells of a worksheet. Before creating a chart, you should prepare it so it can be easily recognizable. Data used on a chart can be made of natural numbers or percentage values. You can also present a series of repeating words and let the chart engine count the occurrences of such words before using them as numbers.

CHARACTERISTICS OF A CHART

A Chart and its Container

A chart cannot reside on its own. It needs a container which is a worksheet. By default, after selecting the cells and starting to create a chart, the chart is created in the same worksheet where the values were selected. If you want, you can put the chart in another worksheet. To do that, right-click the chart and click Move Chart.

A dialog box would display: To put the chart in an existing worksheet, select it in the Object In combo box. To display it in a brand new worksheet, click the New Sheet radio button and specify the name in the top text box.

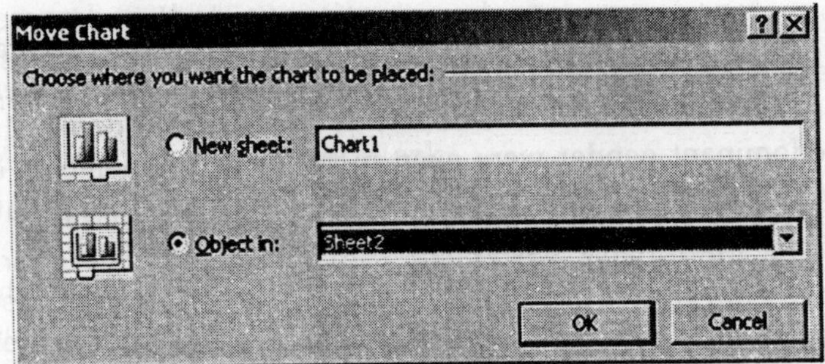

The Sections of a Chart

To present its information, a chart is made of various sections:

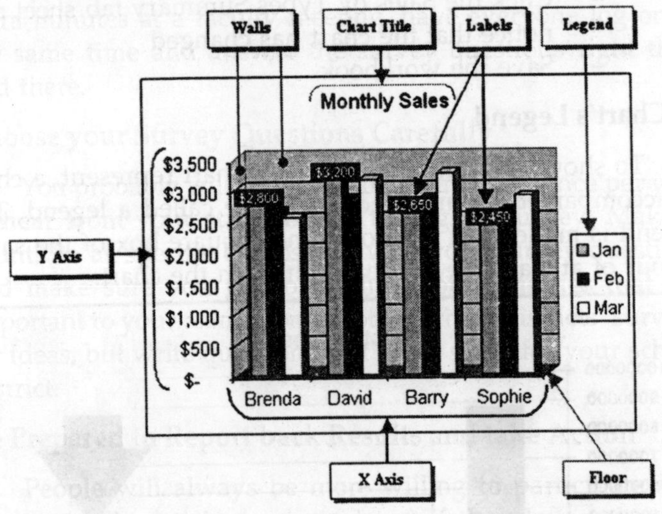

Most or every one of these aspects can be hidden, displayed or changed. To perform any action on these parts, you can right-choose a section or an object on the chart and choose a Format..option.

Editing the Values of a Chart

As mentioned already, to create a chart, you select values from some cells of a worksheet. When analyzing data using a chart, you may want to use "What If" scenarios. For example, if you are viewing the numbers of students per gender in a school and one gender is predominant, you may want to view the tendency if the number of members were the same for both members, if the members of the predominant gender were even more, or if the members of the other gender were predominant. Therefore, during data analysis, you can change the values used by the chart.

To change the values used for a chart, click the appropriate cell on the worksheet and type the desired value. When you do this, the chart would be automatically updated.

- *Practical Learning*: Editing the Values of a Chart
  - Click Sales by Types. Click Cell D6 and type 5
  - Click Cell E6 and type 18
  - Click the Sales by Types Summary tab sheet and notice that the chart has changed
  - Save the workbook

A Chart's Legend

To show what the graphics on a chart represent, a chart is accompanied by an object on a side called a legend. The legend is made of at least one small square box of the same colour of at least one of the graphics on the chart:

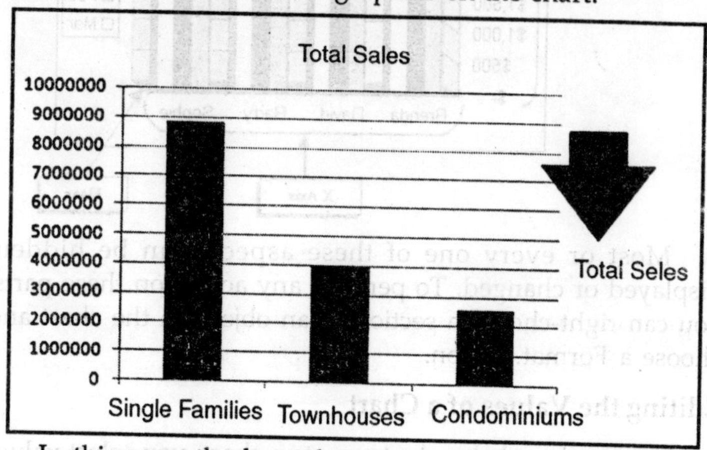

In this case, the legend contains one item labeled Total Sales. Because the legend represents a graphic of the chart, when you make a change on that graphic, the legend is updated. Still, you can change the legend if you want.

To make changes to the legend, on the chart, right-click the legend:

- *To change the font:*
  - Use the buttons on the toolbar
  - Click Font... and use the Font dialog box
- To change other aspects of the legend, click Format Legend... This would display the Format legend dialog box

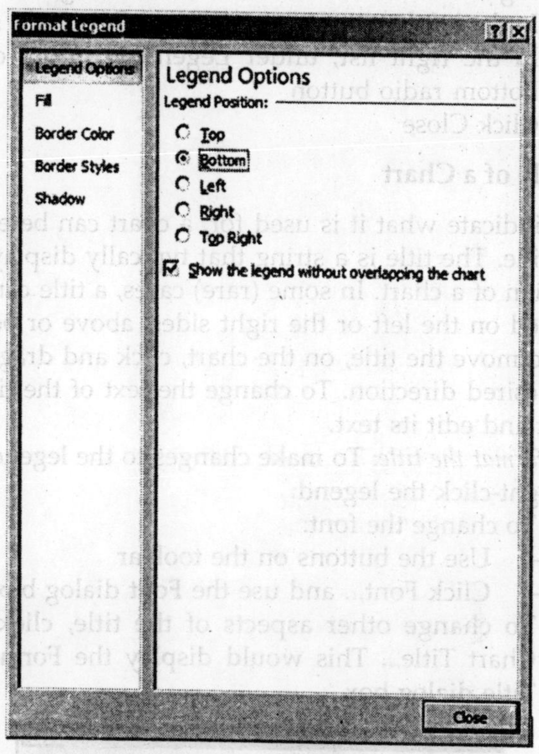

Make the changes, and click Close.

If you do not want to use a legend, you can delete it. To remove the legend:

- Click the legend and press Delete
- Right-click the legend and click Delete

Practical Learning: Using a Chart's Legend

- Right-click Count

- From the toolbar that appears, click the arrow of the Font Name combo box and select Garamond
- Right-click Count again. On the toolbar, click the arrow of the Font Size combo box and select 20
- Click the Bold button
- Click the arrow of the Colour combo box and select Red
- Right-click Count and click Format Legend...
- In the left list, make sure Legend Options is selected. In the right list, under Legend Position, click the Bottom radio button
- Click Close

The Title of a Chart

To indicate what it is used for, a chart can be equipped with a title. The title is a string that typically displays in the top section of a chart. In some (rare) cases, a title can also be positioned on the left or the right sides, above or below the chart. To move the title, on the chart, click and drag the title in the desired direction. To change the text of the title, click inside it and edit its text.

To format the title: To make changes to the legend, on the chart, right-click the legend:

- To change the font:
  - Use the buttons on the toolbar
  - Click Font... and use the Font dialog box
- To change other aspects of the title, click Format Chart Title... This would display the Format Chart Title dialog box

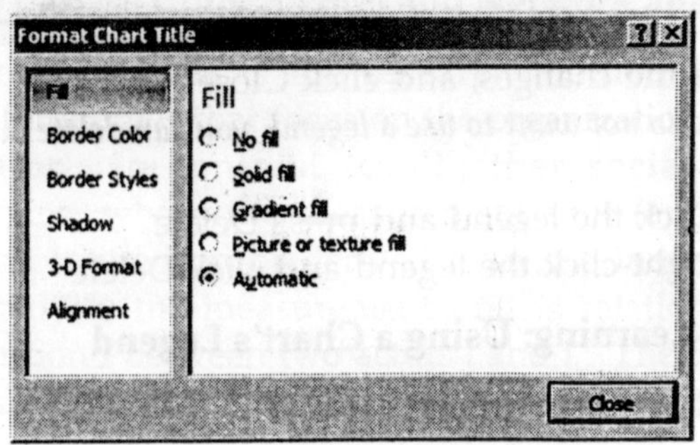

Make the changes, and click Close.

If you do not want to use a title, you can delete it. To remove the title:

- Click the title and press Delete
- Right-click the title and click Delete

Practical Learning: Formatting a Chart's Title

- Click inside the Count label on top and edit it to display Sales Per Type
- Right-click the title
- From the toolbar that appears, click the arrow of the Font Name combo box and select Courier New. Click the arrow of the Font Size combo box and select 28
- Click the arrow of the Colour combo box and select Red
- Save the workbook

Chart Figures

To represent its numbers, a chart draws some geometric figures, depending on the type of chart. These figures can be rectangles, pie slices, triangles, cones, etc. To paint these figures, by default, the chart engine uses some colours from its own list.

You can either change these colours or apply some preset drawings available. You can also design and use any custom picture to paint the chart's shapes.

To format the geometric figures of a chart, you can right-click one of them and click Format Data Series... By default, when you have just created a chart that uses one column for its values, Microsoft Excel applies the same formatting, such as the same colour, to all of its figures. You can keep that common colour or change the colour of each individual shape. To change the formats of a shape, right-click it and click Format Data Point.

Practical Learning: Formatting a Chart's Shapes

- Right-click the left rectangle on the chart and click Format Data Point...

- In the Format Data Point dialog box, in the left list, click Border colour
- In the right list, click Solid Line
- Click the Colour button and click Dark Blue, Text 2, Darker 25%
- In the left list, click Fill
- In the right list, click Gradient Fill
- Click the Preset Colour button and click Ocean
- In the Type combo box, select Rectangle
- Click the Direction button and click the From Center button
- Click Close
- On the chart, right-click the middle rectangle and click Format Data Point...
- In the Format Data Point dialog box, in the left list, click Fill
- In the right list, click Picture or Texture Fill
- Click the Texture button and click Granite
- Click Close
- Save the workbook

Chart's Labels

By default, when a chart is drawn, it is equipped with shapes and a separate legend. If you want, you can display the value of each part and possibly its name close to it. This is done through a label. On a large chart, a label can also be used in the absence of a legend. In fact, you can delete a legend and simply make use of a label.

To add the labels to a chart, right-click a shape on the chart and click Add Data Label. To remove an existing chart label, right-click it and click Delete.

TYPES OF CHARTS

In our introduction to charts, we created one with standing rectangular boxes. This is called a column chart and is only one of the types of charts available. Microsoft Excel (indeed Microsoft Office) provides many other flavours you can use, depending on the type of analysis you want to perform. To select a type of chart, after selecting the cells on a worksheet, on the Ribbon, click Insert. In the Chart section, click one of the buttons to see its options and select from the list. After creating a chart, to change its type, right-click the chart and click Change Chart Type. This would open the Change Chart Type dialog box where you can select a different type.

The Types of Values of a Chart

When you select the cells of a worksheet to create a graph, by default, the application counts the number of occurrences of each value, especially if you select some string-based cells. Depending on the type of chart as we will see in the next few sections, some charts can use regular numbers while some others are better with percentage values. Fortunately, instead of trying to figure out how to perform the calculations yourself, Microsoft Excel can do it for you.

Types of Charts: Column Charts

As we have seen already, a column chart creates vertically standing rectangular boxes. Each box can be used to represent an integral, a decimal, or a percentage value. When creating such a chart, you specify the values to use. Microsoft Exccel determines the highest and the lowest values. When the boxes are drawn, each must fit in the area allocated for the chart. As a consequence, the box that represents the highest value is also the tallest while the box for the lowest value is the shortest. Microsoft Excel draws the other boxes between these extremes but proportionately. Therefore, a column graph is used to compare values in increment.

Double-Column Charts

The classic column chart is made of flat bars that simply illustrate maximal, minimal, and in-between values. One of the options allows you to create a 3-dimensional look of the chart and further accentuate the colours and/or other graphic effects. To enhance an effective analysis, you can create a real 3-D chart that shows data and graphics in perspective.

Another variance of the column chart is to show two columns for each sample value. For example, in our real estate application, imagine that you have the market value for each property and the value the property was sold for, one of the effects of a sale is that some properties would be sold for the same market value, some properties would be sold for a lower value (for example, the seller may want to get rid of the house and be willing to assist the buyer with a down payment and closing cost, thus lowering the price of the house), some other properties could be sold higher than the advertised value (for example, a customer may want to insist on having the house, even at a high price, or too many people could be suddenly interested in the same house, this could raise the price). At the end of the year, when doing an inventory or an evaluation of some sort, you may want to know what houses sold high and which ones sold low.

3-D Column Charts

So far, we have used what are referred to as flat charts. They can be drawn on a 2 dimensional coordinate system. To enhance the appearance of a chart, you can draw it in 3 dimensional coordinate system (x, y, z). If you want to draw 3-D chart, you must select three series of cells. Two of the series should hold categories of values and the other one can hold unique values.

The two series that hold categories of values should have corresponding values so that, a value from one series can have corresponding values in the second series. Here is an example. Imagine that, in a real estate database, you have been selling properties over a period of 1, 2, 3 or more years. The properties sold are categorized as single families, townhouses, and condominiums. Obviously in a particular year, you sell properties of all kinds. On the other hand, each property can have its own value. You can use these three sets of values to create a 3-D chart.

The cone, cylinder, and pyramid charts can be used in the same scenario as the column char. Their 3-D visual effect can enhance the overall analysis of data.

The cylinder chart creates long circular boxes of the same base on both ends. It can be enhanced with good formatted Fill Effects. This chart is suitable for industry, manufacturing analysis, and predictions.

The cone chart is made of a circular base topped by a higher point. When used with various data, the higher values will have the complete cone while the lower values will share portion of the geometric figure. The cone chart should be used with values that can take advantage of its graphing dimensions.

The pyramid chart resembles the cone chart with a difference on their respective bases. Both are constructed the same and can be used in similar scenarios. Or there is too much room on one side. You can rotate the chart. To do this, click one of the borders of the walls of the chart to select its frame. Then click one of the handles on the frame and hold the mouse down. The actual frame of the chart would appear:

You can then rotate the chart in the direction of your choice. You can keep doing this, releasing the mouse to preview, then rotating again, until you get the desired orientation. If you created the chart as one shape (cylinder, cone, or pyramid) but want to use another shape, you can change it.

Bar Charts

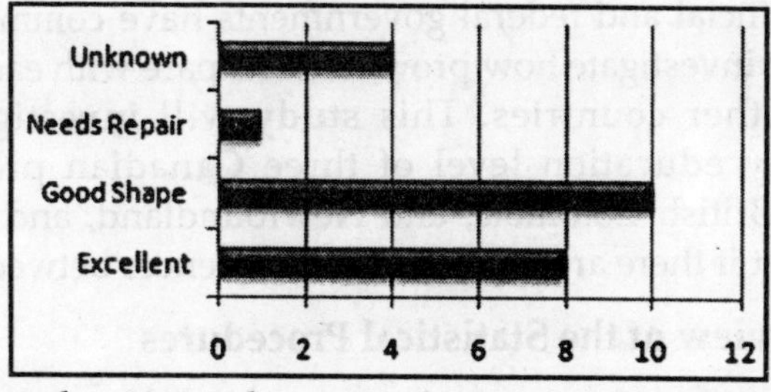

A bar chart uses the same theories and scenarios as the column chart except that its rectangular boxes are horizontal. Like the column chart, the bar chart is used to compare values of the same category on a common scale. You create a bar chart using the same process as the Column, except that you should select the Bar Chart on the Ribbon. As done for the column chart, when specifying the values of a bar chart, use a series of cells that have frequent occurrences of the same values.

ANALYSIS OF VARIANCE

INTRODUCTION

During the last few years in Canada, concern has been growing over the rising provincial and federal deficits, which in no small part, can be linked to the escalating costs of education. However, this does not mean that Canadians as a group have become better educated and are topping the list in international comparison testing. Reports coming from the media have long suggested that as education costs have risen, levels of learning have fallen.

As a result, public outcries have increased, demanding educational inquiries and subsequent reforms to prepare Canadian young people for market demands of the 21st century, and help them to judge the opportunities and choices available to them. All provinces and territories are reviewing their programs and revising their school curricula. The quality and significance of education, along with how well people do in the lower grades, does affect the likelihood that they will go on to pursue a university education.

Provincial and federal governments have commissioned studies to investigate how provinces compare with each other, and to other countries. This study will investigate the university education level of three Canadian provinces: Ontario, British Columbia, and Newfoundland, and attempt to find out if there are any inherent differences between them.

An Overview of the Statistical Procedures

Three random samples of forty degree holders, one sample from each of the three provinces (British Columbia, Newfoundland, and Ontario) were collected along the geographic units of census subdivisions, including cities, towns, and municipalities. Sample data was provided from the 1991 Census, and included two variables: first, that population numbers were limited to only those fifteen years of age and older; and that those studied, had at least one university degree.

Testing will be done based on two separate criteria, namely, the unmodified degree numbers, and the per capita figures, using an alpha level of.05. Two separate null hypotheses and two alternative hypotheses are needed since one F-test will compare the three provinces in the number of people with at least one university degree, and the other will compare the three provinces in degree holders per capita.

For the degrees, the null hypothesis is that there is no difference in the number of people having at least one university degree in the provinces of British Columbia, Ontario, and Newfoundland, while the alternative is that there is a significant difference in the numbers of people having at least one university degree in those three provinces.

For the per capita figures, the null hypothesis is that in the three provinces, there is no difference in the number of people with at least one degree per capita, and the alternative hypothesis states that there is a difference in the number of people with at least one degree per capita in those provinces—at least one province will have more or less degree holders per capita. Since looking at the sample means of each group will not necessarily give a realistic picture of significant difference, we must check the standard deviation to find the variance.

When comparing more than two means simultaneously, analysis of variance (ANOVA) is used "to determine whether any differences among two or more means are greater than would be expected by chance" (Walsh 1990,124). The ANOVA test will provide data on the "statistical significance of a relationship," by examining the ratio of between-group variance/within-group variance, and give a F-score.

The first step is to compare the three provinces in the number of people with at least one university degree, by using the analysis of variance (ANOVA). This results in a F-ratio of 1.587, which is clearly lower than the F-critical value of 3.074.

Thus, we will fail to reject the null hypothesis based on this test, since the difference is not significant enough to say that it is based on anything but chance. Since there are no significant differences among the samples, there is no point in examining the data any further. Simply counting the numbers of degrees that a province has is not enough to give a clear picture.

The figures are inconclusive, but going further and testing for the F-score of per capita figures will provide a more realistic picture of the level of university education within these three test provinces. Therefore, using the per capita figures that were generated by the computer software we can compare the three provinces in degree holders per capita.

This test, results in an F value that clearly exceeds the F-critical value, forcing us to reject the null hypothesis and accept the alternative one, which states that there is a significant difference in the number of people with at least one degree per capita in British Columbia, Newfoundland, and Ontario. What it does not tell us, is the pattern of difference, and therefore, the HSD or Tukey Test must be used to locate from where the difference is coming in the per capita figures. The HSD value in this case is.0263.

The results of that test are significant and suggest patterns of difference. First, the pairwise comparison of the absolute mean values of British Columbia and Newfoundland equal.0022, which is less than the HSD benchmark of.0263. This suggests that those two provinces are not significantly different from each other. Consequently, British Columbia and Newfoundland have similar numbers of people holding at least one degree, per capita.

The second phase of the pairwise comparisons shows a value of.0285 between British Columbia and Ontario, a number that is clearly greater than the HSD figure of.0263. This implies that there is a significant difference in the number of people with at least one degree, per capita, between British Columbia and Ontario, and in fact is reinforced when one looks at the raw mean per capita figures for both provinces. Thirdly, the pairwise calculations between Newfoundland and

Ontario produces the figure.0307, which is also greater than the HSD, demonstrating that a significant difference exists when comparing the number of degree holders per capita within those two provinces, with Ontario having more.

These findings are confirmed when they are compared with the confidence intervals of each province. Since the study is dealing "with sample data, one can state the probability that a parameter falls within a specified range called the confidence interval". We can be 95% certain that the number of people per capita in British Columbia with at least one degree will fall between.0207 and.0568, +/- 1.96 standard errors of the samples estimated.

For Newfoundland. 0217 and .0568, and for Ontario they will fall between.0533 and.0813. As can be seen, there is great overlap between British Columbia's confidence intervals and Newfoundland's, which does confirm what the HSD found. In addition, there is little overlap when comparing the confidence levels of British Columbia and Ontario, and Newfoundland and Ontario. As A. Walsh has written in*Statistics for the Social Sciences: With Computer Applications*:

The objective of social science is to determine the source of this variability. We ask questions like "What is it that 'causes' or accounts for the observed differences between this group of people and that group of people, this individual or that individual".

Interpretations

Why would British Columbia and Newfoundland have similar numbers of university degree holders per capita, and fall far below the Ontario numbers? The major concern, when using analysis of variance is that the samples used must be random.

For instance, if the samples collected in British Columbia were not taken from all over the province, including the highly populated areas of Vancouver and Victoria, where one would expect to find many degree holders living, then the data may not truly reflect a population. In other words, the sample will not be free of bias. The results for British

Columbia are particularly puzzling, (especially when looking at the standard deviation numbers for B. C. per capita) not because they show that the province has less degree holders per capita when compared with Ontario, but that the figures are so low and comparable to Newfoundland's. O'Sullivan and Rassel have pointed out that although random sampling is the goal, "this requirement is frequently violated... [and] the user should not infer a precise meaning from the statistic". Regardless, with caution these statistics may provide some insight into the realm of university education in Canada, and bring about new avenues of research.

The first point to clarify is that education is a provincial responsibility under the Canadian Constitution, and so, educational systems and practices are directly affected by the policies, standards, and budgets that are set out by politicians and bureaucrats in the individual provinces. For many decades, Ontario was Canada's richest province, although the last few years have put Alberta in line for first place. Regardless, Ontario still spends more money per student on education than any other province per capita, and has one of the lowest student/teacher ratios. Ontario is also known as the province that pays their teachers much more than all other provinces.

Does this mean that the best Canadian teachers are flocking to Ontario for monetary reasons, and thus, supplying the students there with obvious benefits? Further research will be needed to answer this question, however, it can be said that lack of money does affect an educational program and deprives students of the diversity and depth of knowledge needed to spark an interest in a university education and propel people into lifelong learning. Education in the lower grades is a fundamental factor.

Newfoundland is considered a 'have not' province, and has never had the large sums of money like Ontario for its education system. Therefore, the quality of education has never been as high in that province. The other factor is that for many years, most of its teachers had no university education at all, with some having one year. There are countless stories of the late Hon. Joseph R. Smallwood, one-time premier of Newfoundland and Father of Confederation, going to tiny outports to handpick future teachers and university professors, when the university opened in the 1960s. He chose the brightest students in an area, but often they had not even finished high school themselves, and many of these are still teaching today. As well, British Columbia's economy has always been fairly stable, but the province has not had huge amounts of money to pour into their school system.

Other possibilities may shed some light upon the provincial differences recorded in this study, including accessibility to a university. For instance, Ontario must have more universities than any other province, while Newfoundland only has one, and British Columbia has few. If the majority of Newfoundlanders live in rural areas, far away from the capital city and Memorial University, many will not think of leaving their friends, family, and familiar surroundings to get a university education, and some cannot. There is also a realization that having a university education in Newfoundland is synonymous with moving to the mainland—something that many Newfoundlanders find quite disturbing.

Certainly, this has affected the level of university education in the province. Incurring a student loan almost demands a move off the island for its repayment, since jobs are very scarce. British Columbia has an extensive network of community colleges, but very few 'universities' to serve its population.

If the prestige of a university education is preferred, but the person is living in northern British Columbia, it reduces the likelihood of that person going to Vancouver or Victoria, or that the move will be a fiscal possibility, with the high rental costs in those areas. Ontario's population does not have the same concerns, as the numbers and quality of its provincial universities is great, which significantly reduces the problems of access, and therefore, attracts more students. These factors are particularly important as the trend is continuing that sees ever more students living at home to keep post-secondary costs down.

Finally, there may be a higher probability that people living in British Columbia and Newfoundland will attend a technical school rather than university. Looking at the numbers employed in natural resource-based industries in these two provinces seems to point in that direction. British Columbia has its farming, forestry, and fishing industries, plus a high concentration of jobs in the movie and music business. Also, construction jobs have been very plentiful in that province for the last decade due in no small part to the influx of the Hong Kong immigrants.

These kinds of jobs require technical training and not scholarly instruction. Likewise, Newfoundland has been a province that has relied heavily on mining, fishing, and tourism-related industries for its economy, which may or may not call for technical training.

Currently, Hibernia is one big employer, with workers getting their training at the local technical college. In addition, drop out rates have always been high in the province, much of which is attributed to apprenticeships in the fishery at an early age. Also, both British Columbia and Newfoundland have very high illiteracy rates, with Newfoundland predictably having the highest illiteracy rate among provinces. William Thorsell, editor-in chief of the*Globe and Mail* has reported that 31 per cent of British Columbia's population is illiterate.

This differs in Ontario, which historically has had a heavy immigrant population coming from cultures that value education, the work ethic, and aspirations of greater educational performance, with a will to succeed in that provinces' economic community. To do that, university educations are more likely to be needed as the numbers of professional and highly skilled positions in corporations, corporate law firms, and head offices of financial institutions warrant. Lastly, it must be remembered that there are particular provinces that people leave in great numbers, and those that attract many new residents every year.

Newfoundlanders have been migrating from the province in large numbers for many decades, living away from their hometowns with little chance of returning. This multitude of Canadians show up in the census figures of their new place of residence. For example, one of the largest populations of Newfoundlanders living outside the province, live in the greater Toronto area, and some of them will have at least one degree or more.

Purpose

The reason for doing an ANOVA is to see if there is any difference between groups on some variable. For example, you might have data on student performance in non-assessed tutorial exercises as well as their final grading. You are interested in seeing if tutorial performance is related to final grade. ANOVA allows you to break up the group according to the grade and then see if performance is different across these grades. ANOVA is available for both parametric (score data) and non-parametric (ranking/ordering) data.

TYPES OF ANOVA

One-way between groups

The example given above is called a one-way between groups model. You are looking at the differences between the groups. There is only one grouping (final grade) which you are using to define the groups. This is the simplest version of ANOVA. This type of ANOVA can also be used to compare variables between different groups-tutorial performance from different intakes.

One-way Repeated Measures

A one way repeated measures ANOVA is used when you have a single group on which you have measured something a few times.

For example, you may have a test of understanding of Classes. You give this test at the beginning of the topic, at the end of the topic and then at the end of the subject.

You would use a one-way repeated measures ANOVA to see if student performance on the test changed over time.

Two-way Between Groups

A two-way between groups ANOVA is used to look at complex groupings. For example, the grades by tutorial analysis could be extended to see if overseas students performed differently to local students.

What you would have from this form of ANOVA is:

- The effect of final grade
- The effect of overseas versus local
- The interaction between final grade and overseas/ local

Each of the main effects are one-way tests. The interaction effect is simply asking "is there any significant difference in performance when you take final grade and overseas/local acting together".

Two-way Repeated Measures

This version of ANOVA simple uses the repeated measures structure and includes an interaction effect. In the example given for one-way between groups, you could add Gender and see if there was any joint effect of gender and time of testing-*i.e.* do males and females differ in the amount they remember/absorb over time.

NON-PARAMETRIC AND PARAMETRIC

ANOVA is available for score or interval data as parametric ANOVA. This is the type of ANOVA you do from the standard menu options in a statistical package. The non-parametric version is usually found under the heading "Nonparametric test". It is used when you have rank or ordered data.

You cannot use parametric ANOVA when you data is below interval measurement. Where you have *categorical* data you do not have an ANOVA method-you would have to use Chi-square which is about interaction rather than about differences between groups.

HOW ITS DONE

What ANOVA looks at is the way groups differ internally versus what the difference is between them.

To take the above example:

- ANOVA calculates the mean for each of the final grading groups (HD, D, Cr, P, N) on the tutorial exercise figure-the Group Means.
- It calculates the mean for all the groups combined-the Overall Mean.
- Then it calculates, within each group, the total deviation of each individual's score from the Group Mean - Within Group Variation.
- Next, it calculates the deviation of each Group Mean from the Overall Mean - Between Group Variation.
- Finally, ANOVA produces the F statistic which is the ratio Between Group Variation to the Within Group Variation.

If the Between Group Variation is significantly greater than the Within Group Variation, then it is likely that there is a statistically significant difference between the groups. The statistical package will tell you if the F ratio is significant or not.

All versions of ANOVA follow these basic principles but the sources of Variation get more complex as the number of groups and the interaction effects increase.

Comments

It is unlikely that you would do an analysis of variance by hand. Except for small data sets it is very time consuming.

AVAILABLE SOFTWARE

- *SPSS*: The ANOVA routines in SPSS are OK for simple one-way analyses. Anything more complicated gets difficult. All statistical packages (SAS, Minitab etc.) provide for ANOVA.
- *Excel*: Excel allows you to ANOVA from the Data Analysis Add-on. The instructions are not good.

ANALYST APPLICATION ANOVA

Analysis of Variance

The ANOVA tasks in the Analyst Application build in complexity. The One-Way ANOVA offers the most basic specification, and the Factorial ANOVA extends the possible model to multiple classification variables.

A Linear Models task provides a more comprehensive set of tools for modeling any number of response variables by both quantitative and class independent variables. Finally, you can use the Repeated Measures task to model correlated responses using a wide range of covariance structures, and you can specify a model with both fixed and random effects using the Mixed Models task.

One-way Analysis of Variance

In the One-Way ANOVA task, you can perform a one-way analysis of variance (ANOVA). This analysis technique is used for experimental data in which there are a continuous response variable and a single independent classification variable. In the One-Way ANOVA task, you can produce a variety of tests, request means comparisons and plots, and specify By group variables. This task is appropriate if you have a single classification variable and the data are balanced.

Factorial Anova

In the Factorial ANOVA task, you can perform a factorial analysis of variance. This analysis technique is used for experimental data in which there are a continuous response variable and one or more independent classification variables. This analysis is appropriate if you have multiple classification variables or a single classification variable with unbalanced data (not all levels have the same number of data points).

Linear Models

In the Linear Models task, you can fit general linear models with the method of least squares. This analysis technique is used for experimental data in which there are a continuous response variable and one or more independent classification variables as well as one or more independent quantitative variables. In the linear models task, you can specify a sophisticated model, produce a variety of tests, and request means comparisons and least squares means. You can also request plots displaying interactions, predicted values, and a number of diagnostic statistics.

Repeated Measures

You can use the Analyst Application to perform repeated measures analysis. Repeated measures refer to multiple measurements on an experimental unit, such as right- and left-eye visual acuity, or measurements made over a period of time, such as blood pressure measured once a week for a month. Questions you want to answer in a repeated measures analysis may be whether treatment influences the response, whether time, the repeated factor, influences the response, and whether there is a treatment by time interaction.

Using the repeated measures task, you can specify a model, repeated effect, subject effect, and covariance structure for the measurements on an individual subject. Your model can incorporate both quantitative and classification variables for main effects, interactions, polynomial terms, and nested effects. The same covariance structure is used for all subjects, and you can select from a wide range of structures, including unstructured, autoregressive(1), and compound symmetry. You can request hypothesis tests, produce least squares means, compute descriptive statistics, and plot means, predicted values, and residuals. The task uses the mixed models approach for analyzing repeated measures.

View an example of a Repeated Measures analysis using the Analyst Application.

Mixed Models

In the Mixed Models task, you can fit linear models that incorporate both fixed and random effects. With random effects, the levels in the factor that are used in the study represent a random sample of a larger set of possible levels.

The distribution of the possible levels of the factor has a mean and a variance. Often, the random effect is that of a blocking factor, such as blocks in a split plot design. Using the Mixed Models task, you can specify fixed and random effects that include interactions, polynomials, and nesting.

You can produce tests concerning fixed effects and variance components, request a specific estimation method, and compute least squares means and pairwise differences. In addition, you can output predicted values as well as plot means, predictions, and residuals.

DATA ANALYSIS TOOLS

EXCEL FOR STATISTICAL DATA ANALYSIS

At A Glance

We used Excel to do some basic data analysis tasks to see whether it is a reasonable alternative to using a statistical package for the same tasks. We concluded that Excel is a poor choice for statistical analysis beyond textbook examples, the simplest descriptive statistics, or for more than a very few columns.

The problems we encountered that led to this conclusion are in four general areas:

- Missing values are handled inconsistently, and sometimes incorrectly.
- Data organization differs according to analysis, forcing you to reorganize your data in many ways if you want to do many different analyses.
- Many analyses can only be done on one column at a time, making it inconvenient to do the same analysis on many columns.
- Output is poorly organized, sometimes inadequately labeled, and there is no record of how an analysis was accomplished.

Excel is convenient for data entry, and for quickly manipulating rows and columns prior to statistical analysis. However when you are ready to do the statistical analysis, we recommend the use of a statistical package such as SAS, SPSS, Stata, Systat or Minitab.

Introduction

Excel is probably the most commonly used spreadsheet for PCs. Newly purchased computers often arrive with Excel already loaded. It is easily used to do a variety of calculations, includes a collection of statistical functions, and a Data Analysis ToolPak. As a result, if you suddenly find you need to do some statistical analysis, you may turn to it as the obvious choice. We decided to do some testing to see how well Excel would serve as a Data Analysis application.

To present the results, we will use a small example. The data for this example is fictitious. It was chosen to have two categorical and two continuous variables, so that we could test a variety of basic statistical techniques. Since almost all real data sets have at least a few missing data points, and since the ability to deal with missing data correctly is one of the features that we take for granted in a statistical analysis package, we introduced two empty cells in the data:

Treatment	Outcome	X	Y
1	1	10.2	9.9
1	1	9.7	
2	1	10.4	10.2
1	2	9.8	9.7
2	1	10.3	10.1
1	2	9.6	9.4
2	1	10.6	10.3
1	2	9.9	9.5
2	2	10.1	10
2	2		10.2

Each row of the spreadsheet represents a subject. The first subject received Treatment 1, and had Outcome 1. X and Y are the values of two measurements on each subject. We were unable to get a measurement for Y on the second subject, or on X for the last subject, so these cells are blank. The subjects are entered in the order that the data became available, so the data is not ordered in any particular way.

We used this data to do some simple analyses and compared the results with a standard statistical package. The comparison considered the accuracy of the results as well as the ease with which the interface could be used for bigger data sets-*i.e.* more columns. We used SPSS as the standard, though any of the statistical packages OIT supports would do equally well for this purpose. In this article when we say "a statistical package," we mean SPSS, SAS, STATA, SYSTAT, or Minitab. Most of Excel's statistical procedures are part of the Data Analysis tool pack, which is in the Tools menu. It includes a variety of choices including simple descriptive statistics, t-tests, correlations, 1 or 2-way analysis of variance, regression, etc. If you do not have a Data Analysis item on the Tools menu, you need to install the Data Analysis ToolPak. Search in Help for "Data Analysis Tools" for instructions on loading the ToolPak.

Two other Excel features are useful for certain analyses, but the Data Analysis tool pack is the only one that provides reasonably complete tests of statistical significance. Pivot Table in the Data menu can be used to generate summary tables of means, standard deviations, counts, etc. Also, you could use functions to generate some statistical measures, such as a correlation coefficient. Functions generate a single number, so using functions you will likely have to combine bits and pieces to get what you want. Even so, you may not be able to generate all the parts you need for a complete analysis.

Unless otherwise stated, all statistical tests using Excel were done with the Data Analysis ToolPak.

In order to check a variety of statistical tests, we chose the following tasks:

- Get means and standard deviations of X and Y for the entire group, and for each treatment group.
- Get the correlation between X and Y.
- Do a two sample t-test to test whether the two treatment groups differ on X and Y.
- Do a paired t-test to test whether X and Y are statistically different from each other.

- Compare the number of subjects with each outcome by treatment group, using a chi-squared test.

All of these tasks are routine for a data set of this nature, and all of them could be easily done using any of the aobve listed statistical packages.

General Issues

- *Enable the Analysis ToolPak*: The Data Analysis ToolPak is not installed with the standard Excel setup. Look in the Tools menu. If you do not have a Data Analysis item, you will need to install the Data Analysis tools. Search Help for "Data Analysis Tools" for instructions.
- *Missing Values:* A blank cell is the only way for Excel to deal with missing data. If you have any other missing value codes, you will need to change them to blanks.
- *Data Arrangement*: Different analyses require the data to be arranged in various ways. If you plan on a variety of different tests, there may not be a single arrangement that will work. You will probably need to rearrange the data several ways to get everything you need.
- *Dialog Boxes*: Choose Tools/Data Analysis, and select the kind of analysis you want to do. The typical dialog box will have the following items:
- *Input Range:* Type the upper left and lower right corner cells. *e.g.* A1:B100. You can only choose adjacent rows and columns. Unless there is a checkbox for grouping data by rows or columns (and there usually is not), all the data is considered as one glop.
- *Labels:* There is sometimes a box you can check off to indicate that the first row of your sheet contains labels. If you have labels in the first row, check this box, and your output MAY be labeled with your label. Then again, it may not.
- *Output location:* New Sheet is the default. Or, type

in the cell address of the upper left corner of where you want to place the output in the current sheet. New Worksheet is another option, which I have not tried. Ramifications of this choice are discussed below. Other items, depending on the analysis.

- *Output location*: The output from each analysis can go to a new sheet within your current Excel file (this is the default), or you can place it within the current sheet by specifying the upper left corner cell where you want it placed. Either way is a bit of a nuisance. If each output is in a new sheet, you end up with lots of sheets, each with a small bit of output. If you place them in the current sheet, you need to place them appropriately; leave room for adding comments and labels; changes you need to make to format one output properly may affect another output adversely.
  - *Example*: Output from Descriptives has a column of labels such as Standard Deviation, Standard Error, etc. You will want to make this column wide in order to be able to read the labels. But if a simple Frequency output is right underneath, then the column displaying the values being counted, which may just contain small integers, will also be wide.

Results of Analyses

- *Descriptive Statistics:* The quickest way to get means and standard deviations for a entire group is using Descriptives in the Data Analysis tools. You can choose several adjacent columns for the Input Range (in this case the X and Y columns), and each column is analysed separately. The labels in the first row are used to label the output, and the empty cells are ignored. If you have more, non-adjacent columns you need to analyse, you will have to repeat the process for each group of contiguous columns. The procedure is straightforward, can manage many columns reasonably efficiently, and empty cells are treated properly. To get the means and standard deviations of X and Y for each treatment group requires the use of Pivot Tables (unless you want to rearrange the data sheet to separate the two groups). After selecting the (contiguous) data range, in the Pivot Table Wizard's Layout option, drag Treatment to the Row variable area, and X to the Data area. Double click on "Count of X" in the Data area, and change it to Average. Drag X into the Data box again, and this time change Count to StdDev. Finally, drag X in one more time, leaving it as Count of X. This will give us the Average, standard deviation and number of observations in each treatment group for X. Do the same for Y, so we will get the average, standard deviation and number of observations for Y also. This will put a total of six items in the Data box (three for X and three for Y). As you can see, if you want to get a variety of descriptive statistics for several variables, the process will get tedious. A statistical package lets you choose as many variables as you wish for descriptive statistics, whether or not they are contiguous. You can get the descriptive statistics for all the subjects together, or broken down by a categorical variable such as treatment. You can select the statistics you want to see once, and it will apply to all variables chosen.

- *Correlations:* Using the Data Analysis tools, the dialog for correlations is much like the one for descriptives-you can choose several contiguous columns, and get an output matrix of all pairs of correlations. Empty cells are ignored appropriately. The output does NOT include the number of pairs of data points used to compute each correlation (which can vary, depending on where you have missing data), and does not indicate whether any of the correlations are statistically significant. If you want correlations on non-contiguous columns, you would either have to include the intervening columns, or copy the desired columns to a contiguous location. A statistical package would permit you to choose non-contiguous columns for your correlations. The output would tell you how many pairs of data points were used to compute each correlation, and which correlations are statistically significant.

- *Two-Sample T-test:* This test can be used to check whether the two treatment groups differ on the values of either X or Y. In order to do the test you need to enter a cell range for each group. Since the data were not entered by treatment group, we first need to sort the rows by treatment. Be sure to take all the other columns along with treatment, so that the data for each subject remains intact. After the data is sorted, you can enter the range of cells containing the X measurements for each treatment. Do not include the row with the labels, because the second group does not have a label row. Therefore your output will not be labeled to indicate that this output is for X. If you want the output labeled, you have to copy the cells corresponding to the second group to a separate column, and enter a row with a label for the second group. If you also want to do the t-test for the Y measurements, you'll need to repeat the process. The empty cells are ignored, and other than the problems with labeling the output, the results are correct. A statistical package would do this task without any need to sort the data or copy it to another column, and the output would always be properly labeled to the extent that you provide labels for your variables and treatment groups. It would also allow you to choose more than one variable at a time for the t-test (*e.g.* X and Y).
- *Paired t-test:* The paired t-test is a method for testing whether the difference between two measurements on the same subject is significantly different from 0. In this example, we wish to test the difference

between X and Y measured on the same subject. The important feature of this test is that it compares the measurements within each subject. If you scan the X and Y columns separately, they do not look obviously different. But if you look at each X-Y pair, you will notice that in every case, X is greater than Y. The paired t-test should be sensitive to this difference. In the two cases where either X or Y is missing, it is not possible to compare the two measures on a subject. Hence, only 8 rows are usable for the paired t-test. When you run the paired t-test on this data, you get a t-statistic of 0.09, with a 2-tail probability of 0.93. The test does not find any significant difference between X and Y. Looking at the output more carefully, we notice that it says there are 9 observations. As noted above, there should only be 8. It appears that Excel has failed to exclude the observations that did not have both X and Y measurements. To get the correct results copy X and Y to two new columns and remove the data in the cells that have no value for the other measure. Now re-run the paired t-test. This time the t-statistic is 6.14817 with a 2-tail probability of 0.000468. The conclusion is completely different! Of course, this is an extreme example. But the point is that Excel does not calculate the paired t-test correctly when some observations have one of the measurements but not the other. Although it is possible to get the correct result, you would have no reason to suspect the results you get unless you are sufficiently alert to notice that the number of observations is wrong. There is nothing in online help that would warn you about this issue. Interestingly, there is also a TTEST function, which gives the correct results for this example. Apparently the functions and the Data Analysis tools are not consistent in how they deal with missing cells. Nevertheless, I cannot recommend the use of functions in preference to the Data Analysis tools, because the result of using a function is a single number-in this case, the 2-tail probability of the t-statistic. The function does not give you the t-statistic itself, the degrees of freedom, or any number of other items that you would want to see if you were doing a statistical test. A statistical packages will correctly exclude the cases with one of the measurements missing, and will provide all the supporting statistics you need to interpret the output.

- *Crosstabulation and Chi-Squared Test of Independence:* Our final task is to count the two outcomes in each treatment group, and use a chi-square test of independence to test for a relationship between treatment and outcome. In order to count the outcomes by treatment group, you need to use Pivot Tables. In the Pivot Table Wizard's Layout option, drag Treatment to Row, Outcome to Column and also to Data. The Data area should say "Count of Outcome" – if not, double-click on it and select "Count". If you want percents, double-click "Count of Outcome", and click Options; in the "Show Data As" box which appears, select "% of row". If you want both counts and percents, you can drag the same variable into the Data area twice, and use it once for counts and once for percents. Getting the chi-square test is not so simple, however. It is only available as a function, and the input needed for the function is the observed counts in each combination of treatment and outcome (which you have in your pivot table), and the expected counts in each combination. Expected counts? What are they? How do you get them? If you have sufficient statistical background to know how to calculate the expected counts, and can do Excel calculations using relative and absolute cell addresses, you should be able to navigate through this. If not, you're out of luck. Assuming that you surmounted the problem of expected counts, you can use the Chitest function to get the probability of observing a chi-square value bigger than the one for this table. Again, since we are using functions, you do not get many other necessary pieces of the calculation, notably the value of the chi-square statistic or its degrees of freedom. No statistical package would require you to provide the expected values before computing a chi-square test of indepencence. Further, the results would always include the chi-square statistic and its degrees of freedom, as well as its probability. Often you will get some additional statistics as well.

Additional Analyses

The remaining analyses were not done on this data set, but some comments about them are included for completeness.

- *Simple Frequencies:* You can use Pivot Tables to get simple frequencies. (Crosstabulations for more about how to get Pivot Tables.) Using Pivot Tables, each column is considered a separate variable, and labels in row 1 will appear on the output. You can only do one variable at a time. Another possibility is to use the Frequencies function. The main advantage of this method is that once you have defined the frequencies function for one column, you can use Copy/Paste to get it for other columns. First, you will need to enter a column with the values you want counted (bins). If you intend to do the frequencies for many columns, be sure to enter values for the column with the most categories. *e.g.,* if 3 columns have values of 1 or 2, and the fourth has values of 1,2,3,4, you will need to enter the bin values as 1,2,3,4. Now select enough empty cells in one column to store the results-4 in this example, even if the current column only has 2 values. Next choose Insert/ Function/Statistical/Frequencies on the menu. Fill in the input range for the first column you want to count using relative addresses (*e.g.* A1:A100). Fill in the Bin Range using the absolute addresses of the locations where you entered the values to be counted Click Finish. Note the box above the column headings of the sheet, where the formula is displayed. It start with "= FREQUENCIES(". Place the cursor to the left of the = sign in the formula, and press Ctrl-Shift-Enter. The frequency counts now appear in the cells you selected. To get the frequency counts of other columns, select the cells with the frequencies in them, and choose Edit/Copy on the menu. If the next column you want to count is one column to the right of the previous one, select the cell to the right of the first frequency cell, and choose Edit/Paste (ctrl-V). Continue moving to the right and pasting for each column you want to count. Each time you move one column to the right of the original frequency cells, the column to be counted is shifted right from the first column you counted. If you want percents as well, you'll have to use the Sum function to compute the sum of the frequencies, and define the formula to get the per cent for one cell. Select the cell to store the first per cent, and type the formula into the formula box at the top of the sheet-*e.g.* = N1*100/N$5- where N1 is the cell with the frequency for the first category, and N5 is the cell with the sum of the frequencies. Use Copy/Paste to get the formula for the remaining cells of the first column. Once you have the percents for one column, you can Copy/ Paste them to the other columns. You'll need to be careful about the use of relative and absolute addresses! In the example above, we used N$5 for the denominator, so when we copy the formula down to the next frequency on the same column, it will still look for the sum in row 5; but when we copy the formula right to another column, it will shift to the frequencies in the next column. Finally, you can use Histogram on the Data Analysis menu. You can only do one variable at a time. As with the

Frequencies function, you must enter a column with "bin" boundaries. To count the number of occurrences of 1 and 2, you need to enter 0,1,2 in three adjacent cells, and give the range of these three cells as the Bins on the dialog box. The output is not labeled with any labels you may have in row 1, nor even with the column letter. If you do frequencies on lots of variables, you will have difficulty knowing which frequency belongs to which column of data.

- *Linear Regression:* Since regression is one of the more frequently used statistical analyses, we tried it out even though we did not do a regression analysis for this example. The Regression procedure in the Data Analysis tools lets you choose one column as the dependent variable, and a set of contiguous columns for the independents. However, it does not tolerate any empty cells anywhere in the input ranges, and you are limited to 16 independent variables. Therefore, if you have any empty cells, you will need to copy all the columns involved in the regression to new columns, and delete any rows that contain any empty cells. Large models, with more than 16 predictors, cannot be done at all.
- *Analysis of Variance*: In general, the Excel's ANOVA features are limited to a few special cases rarely found outside textbooks, and require lots of data re-arrangements.
- *One-way ANOVA*: Data must be arranged in separate and adjacent columns (or rows) for each group. Clearly, this is not conducive to doing 1-ways on more than one grouping. If you have labels in row 1, the output will use the labels.
- *Two-Factor ANOVA Without Replication:* This only does the case with one observation per cell (*i.e.* no Within Cell error term). The input range is a rectangular arrangement of cells, with rows representing levels of one factor, columns the levels

of the other factor, and the cell contents the one value in that cell.

- *Two-Factor ANOVA with Replicates*: This does a two-way ANOVA with equal cell sizes. Input must be a rectangular region with columns representing the levels of one factor, and rows representing replicates within levels of the other factor. The input range MUST also include an additional row at the top, and column on the left, with labels indicating the factors. However, these labels are not used to label the resulting ANOVA table. Click Help on the ANOVA dialog for a picture of what the input range must look like.

Requesting Many Analyses

If you had a variety of different statistical procedures that you wanted to perform on your data, you would almost certainly find yourself doing a lot of sorting, rearranging, copying and pasting of your data. This is because each procedure requires that the data be arranged in a particular way, often different from the way another procedure wants the data arranged. In our small test, we had to sort the rows in order to do the t-test, and copy some cells in order to get labels for the output.

We had to clear the contents of some cells in order to get the correct paired t-test, but did not want those cells cleared for some other test. And we were only doing five tasks. It does not get better when you try to do more. There is no single arrangement of the data that would allow you to do many different analyses without making many different copies of the data. The need to manipulate the data in many ways greatly increases the chance of introducing errors.

Using a statistical program, the data would normally be arranged with the rows representing the subjects, and the columns representing variables (as they are in our sample data). With this arrangement you can do any of the analyses discussed here, and many others as well, without having to sort or rearrange your data in any way. Only much more complex analyses, beyond the capabilities of Excel and the scope of this article would require data rearrangement.

Working with Many Columns

What if your data had not 4, but 40 columns, with a mix of categorical and continuous measures? How easily do the above procedures scale to a larger problem?

At best, some of the statistical procedures can accept multiple contiguous columns for input, and interpret each column as a different measure. The descriptives and correlations procedures are of this type, so you can request descriptive statistics or correlations for a large number of continuous variables, as long as they are entered in adjacent columns. If they are not adjacent, you need to rearrange columns or use copy and paste to make them adjacent. Many procedures, however, can only be applied to one column at a time. T-tests (either independent or paired), simple frequency counts, the chi-square test of independence, and many other procedures are in this class. This would become a serious drawback if you had more than a handful of columns, even if you use cut and paste or macros to reduce the work. In addition to having to repeat the request many times, you have to decide where to store the results of each, and make sure it is properly labeled so you can easily locate and identify each output. Finally, Excel does not give you a log or other record to track what you have done. This can be a serious drawback if you want to be able to repeat the same (or similar) analysis in the future, or even if you've simply forgotten what you've already done. Using a statistical package, you can request a test for as many variables as you need at once. Each one will be properly labeled and arranged in the output, so there is no confusion as to what's what. You can also expect to get a log, and often a set of commands as well, which can be used to document your work or to repeat an analysis without having to go through all the steps again.

Summary

Although Excel is a fine spreadsheet, it is not a statistical data analysis package. In all fairness, it was never intended to be one. Keep in mind that the Data Analysis ToolPak is an "add-in"-an extra feature that enables you to do a few quick calculations. So it should not be surprising that that is just what it is good for-a few quick calculations.

If you attempt to use it for more extensive analyses, you will encounter difficulties due to any or all of the following limitations:

- Potential problems with analyses involving missing data. These can be insidious, in that the unwary user is unlikely to realise that anything is wrong.
- Lack of flexibility in analyses that can be done due to its expectations regarding the arrangement of data. This results in the need to cut/paste/sort/and otherwise rearrange the data sheet in various ways, increasing the likelyhood of errors.
- Output scattered in many different worksheets, or all over one worksheet, which you must take responsibility for arranging in a sensible way.
- Output may be incomplete or may not be properly labeled, increasing possibility of misidentifying output.
- Need to repeat requests for the some analyses multiple times in order to run it for multiple variables, or to request multiple options.
- Need to do some things by defining your own functions/formulae, with its attendant risk of errors.
- No record of what you did to generate your results, making it difficult to document your analysis, or to repeat it at a later time, should that be necessary.

If you have more than about 10 or 12 columns, and/or want to do anything beyond descriptive statistics and perhaps correlations, you should be using a statistical package. There are several suitable ones available by site license through OIT, or you can use them in any of the OIT PC labs. If you have Excel on your own PC, and don't want to pay for a statistical program, by all means use Excel to enter the data (with rows representing the subjects, and columns for the variables). All the mentioned statistical packages can read Excel files, so you can do the (time-consuming) data entry at home, and go to the labs to do the analysis.

BASIC STATISTICS FOR ENVIRONMENTAL RESEARCH

Descriptive Statistics

Use of Excel:

- Learn some basic skills of using Excel (also see the demonstration)
- Carefully check your completed dataset with the hard copy of the data sheet
- Check if you can find Tool/Data Analysis: if yes, that's fine; if not, then go to Tool/Add-Ins and then select "Analysis ToolPak"/click OK
- All mathematical or statistical function can be found by clicking "*fx*"
- Understand the data format
- Sort all data by sex: highlight all data; then click on Data/Sort/by "M = 0, F = 1"/finally Click OK
- How many males and females?
- Calculate mean, standard deviation, 95% confidence interval, median, Q1 and Q3 for age, weight and length data. For example, we need to perform Descriptive Statistics on all height measurements so that we can check where the mean and median are located on the histogram. Tool/Data analysis/ Descriptive Statistics/OK/Input the height range/Tick group by columns/Tick New Worksheet Ply/Tick Summary statistics and Confidence level of mean/OK; then label the output worksheet.
- Construct and inspect the frequency histogram for height of All students: Tool/Data analysis/histogram/ input the male height by using your mouse to highlight all male height/Select New Worksheet Ply/ Tick Chart Output/Click OK, label the output worksheet. Does the data follow a normal distribution?

- Repeat Step 7, but add a bin range 150, 153, 156....177. Is there any different?
- Repeat Step 7, but add a bin range 150, 154, 158....178.
- Give some brief accounts based on visual comparisons of the histograms between two sexes (for age, height and weight data).

Use of SPSS

- Open the SPSS window
- Cut and Paste your class data onto a SPSS data sheet
- Name and Classify the column and data type
- For example, for examining All height data: Analyse/ Descriptive statistics/Frequencies/Click on height and put it into the other box....
- To select data for analysis *e.g.* female height: Data/ Select cases.....

Simple Chi-square Test (Using Excel)

- For the results of questionnaire interview, 113 people agree and 146 people disagree on a subject. Test if the results were caused by chance.
- CHITEST (actual dataset, expected dataset) = p value for actual Chi-square
- CHIINV (p value, DF) = Chi-square value; DF = K –1 where K is no. of categories
- Critical Chi-square = CHIINV (0.05, DF)
- Try if (a) 150 agree and 160 disagree and (b) 300 agree and 50 disagree

Two Sample Comparison

Use of Excel

Test whether male is teller than female, *i.e.* test Ho: male height <= female height

Checking normality

- Checking homogeneity of variance
- To test the homogeneity of variance of individual height between sex (Ho: equal variances), we need

to perform a F test. Tool/Data analysis/F-test Two-Sample for Variances/OK/Input male height into variable 1 Range and female height into Variable 2 Range/tick New Worksheet Ply/OK, Label the output worksheet as F-test-01. As you can see the F value is less than 1 that is totally wrong. Please also be noted that Excel only can provide one-tailed critical value of F. For 2-tailed critical value, we need to consult a F-table (Table B4 in Zar 1999).

- Therefore, repeat step 11 again, but now put female height into Variable 1 Range and male data as 2 Range. It is because Excel always use the Variable 1 for the larger variance (*i.e.* 2 Range for the denominator). *e.g.* F critical value at 0.05 (1), 55, 32 = 1.724 and at 0.05 (2), 50, 30 = 1.97 (Table B4). We use 1-tailed test here, if calculated F value for our data (1.415) is less than the critical value, then we accept Ho. And we can proceed to a parametric 2-samples t-test assuming equal variances.
- Parametric two-sample t test
- T test to test Ho: male height £ female height and Ha: male height > female height (*i.e.* 1 tailed test); Tool/Data analysis/t-test: Two Sample Assuming Equal Variances/OK/Input the male data into Variable 1 Range and female into Variable 2 Range (Unlike F-test, it doesn't a matter which box you used)/Hypothesis mean difference = 0/tick New Worksheet ply. Click OK. Label the output page as "t-test". With a priory information that males are generally higher than females, we use a 1-tailed test here. *e.g.* If the sample t value = 12.127 is much greater than the critical t value at 0.05(2), 87 = 1.663 ($p < 0.001$), then we reject the Ho and infer that the height of males is significantly higher than females.
- One sample t-test using SPSS (Excel cannot perform this task!)
- Copy the combined class "age" data from Excel and paste them onto the spreadsheet of SPSS. Label the

column as "age" and set Measure as Scale.

- Go to Analyse ® Compare Means ® One Sample T-Test ® Click on the arrow (to include your "age" data for the analysis) ® Enter Test Value as 22 ® click OK
- Check the t value and p value and draw your conclusion.
- Checking homogeneity of variance and compare two samples using t-test (1-tailed)

Excel only:

- Perform a F-test first to check the homogeneity of variance – same as above.
- Back to the data sheet. Tools ® Data Analysis ® T-Test: Two-Sample Assuming Equal Variances ® select the variable 1 and variable 2 (sequence is not important) Input Hypothesized Mean Difference = 0 ® select New Worksheet Ply ® OK
- Based on 1-tailed critical t value and the p value, you should be able to reject or accept the null hypothesis.
- Non-Parametric Test for comparing two samples- Mann-Whiney test (2-tailed test)

SPSS Only:

- Open a New Spreadsheet in SPSS. Try in fourteen 1 and seventeen 2 in the first column. Label this column as "Group" and set Measure as Nominal.
- Copy and paste the "HKCEE female" dataset onto second column, right next to "1" and do the same for "HKCEE male" dataset but next to "2" in the same column. Label this column and set Measure as Scale.
- Go to Analyse ® Nonparametric Tests ® 2 Independent-Samples ® Select Moist and Click on the arrow (to include your "Moist" data for the analysis) ® Select "Group" as the Grouping Variable and then click on Define Group ® Enter 1 and 2, respectively ® click Mann-Whiney test ® OK
- Based on the p value [*i.e.* Asymp. Sig. (2-tailed)], you should be able to reject or accept the null hypothesis. Remember: SPSS only can deal with 2-tailed test, so

you will need to use the Mann-Whiney table for 1-tailed test in order to infer whether accept or reject the 1-tailed null hypothesis.

Paired Two Sample t-Test

Excel:

- Tools ® Data Analysis ® T-Test: Paired Two-Sample for Means ® select the variable 1 (hind leg) and variable 2 (fore leg) Input Hypothesized Mean Difference = 0 ® select New Worksheet Ply ® OK
- Based on 2-tailed critical t value and the p value, you should be able to reject or accept the null hypothesis.

SPSS:

- Open a New Spreadsheet in SPSS. Try in ten 1 and ten 2 in the first column. Label this column as "Group" and set Measure as Nominal.
- Copy and paste both the "hind leg" and "fore leg" datasets onto first and second columns, respectively. Label them as length1 and length2, and set Measure as Scale.
- Go to Analyse ® Compare Means ® Paired-Samples T-Test ® Select both dataset and Click on the arrow ® click OK
- Based on the p value [*i.e.* Sig. (2-tailed)], you should be able to reject or accept the null hypothesis. Remember: SPSS only can deal with 2-tailed test, so you will need to use the t-table for 1-tailed test in order to infer whether accept or reject the 1-tailed null hypothesis.

Non-parametric Wilcoxon's test (Paired samples)

SPSS Only (stay in the same spreadsheet):

- Using the same data and assuming the data "Paired t-test (2-tailed test)" failed the normality test based on the differences between the two groups.
- Go to Analyse ® Nonparametric Tests ® 2 Related-Samples ® Select both datasets and Click on the arrow ® Select "Wilcoxon's test" ® OK

- Based on the p value [*i.e.* Asymp. Sig. (2-tailed)], you should be able to reject or accept the null hypothesis. Remember: SPSS only can deal with 2-tailed test, so you will need to use the Wilcoxon's table for 1-tailed test in order to infer whether accept or reject the 1-tailed null hypothesis

Correlation

Using the given Class Data:

- Calculate body mass index (BMI) = weight in kg/ (height in cm/100)^2
- Select datasets age, sex, height, weight and BMI plus mathematic score and expected course difficulty to SPSS and label the column accordingly.
- Does height increase with age? Ho: $r = 0$ or $r < 0$; Ha: $r > 1$
  a. Analyse ⇒ Correlate ⇒ Bivariate ⇒ Select age and height data ⇒ Pearson ⇒ One-tailed ⇒ OK ⇒ Interpret the results
  b. Graph ⇒ Scatter ⇒ Simple ⇒ Define ⇒ age as x-axis and height as y-axis ⇒ OK
- Multiple correlation tests for age, height, weight and BMI
  a. Analyse ⇒ Correlate ⇒ Bivariate ⇒ Select all four datasets ⇒ Pearson ⇒ Two-tailed ⇒ OK ⇒ Interpret the results
  b. Graph ⇒ Scatter ⇒ Matrix ⇒ Define ⇒ Select all four datasets ⇒ OK
- *Example 4(a) in Excel*: Pearson vs. Spearmen correlation tests. See demonstration.
- Using SPSS, test the correlation between mathematic score and expected course difficulty
  a. Analyse ⇒ Correlate ⇒ Bivariate ⇒ Select both datasets ⇒ Spearmen ⇒ Two-tailed ⇒ OK ⇒ Interpret the results
- *Try the Example in Excel*: Which method(s) is/are comparable to the standard method?
- Try the Example in Excel

Regression Analysis (*e.g.* fisheries Management)

- If we would like to predict fish weight using length, we need to establish a linear regression model. First, plot length (x-axis) against weight (y-axis) under Excel environment. Secondly, "add trend" onto the data point with the equation and r^2 value. It is a curve rather than a straight-line.
- Thus, we need to log-transform both parameters. Plot the graph using the transformed data. Compare the curve shape and r^2 value with those without transformation.
- Run a regression analysis using the Analysis Tool in Excel.
- Test the relationship between body weight (x-axis) and liver weight (y-axis) by repeating the above procedures.

One-way ANOVA

Excel – Data Analysis:

- Plot a bar chart with the mean and standard deviation
- Run F-test between the larger and smaller variances
- If passed the F-test, proceed to run the "ANOVA: Single Factor"
- For example 1, Input all data and Group by Columns; then OK
- Interpret the results

SPSS:

- *Column* 1: diets – 1, 2, 3 and 4
- *Column* 2: shell length data
- Analyse ⇒ Compare Means ⇒ One-way ANOVA ⇒ shell length as Dependent Variable ⇒ Diets as Fixed Factor ⇒ click Post Hoc tests ⇒ click Tukey ⇒ Continue ⇒ OK
- *OR*: Analyse ⇒ General Linear Model ⇒ Univariate ⇒ shell length as Dependent Variable ⇒ Diets as Fixed Factor ⇒ click Post Hoc tests ⇒ click Tukey ⇒ Continue ⇒ OK
- Interpret the results

Kruskal-Wallis test

SPSS

- *Column* 1: diets – 1, 2, 3 and 4
- *Column* 2: shell length data
- Analyse ⇒ Nonparametric Tests ⇒ K-independent samples ⇒ shell length as Test Variable ⇒ Diets as Grouping Variable (1, 4) ⇒ Kruskal-Wallis H ⇒ OK
- Interpret the results

SPSS FOR DATA PROCESSING

SPSS (Statistical Package for the Social Scientists) is a data management and statistical analysis tool which has a very versatile data processing capability.

Why use SPSS

In the context of small-scale learning and teaching evaluation you are only likely to use a very limited range of the full capability of SPSS. But it can be useful for: Electronically storing questionnaire data. Data is stored in a spreadsheet-like table similar to that of Microsoft Excel. Generating routine descriptive statistical data for question responses, such as frequency counts of closed questions, distribution of multiple-choice question responses etc. Creating graphical presentations of questionnaire data for reporting, presentations or publications. Exploring relationships between responses to different questions.

Collating open question responses. [But there is a 255 character limit on data fields, so the package may not be suitable for collating long qualitative text responses. In this case it might be more suitable to collate qualitative comments using a word processing package.]

Using SPSS

The software can be obtained from Corporate Information and Computing Services (CICS). Training courses are available and can be booked through the Learning Management System, accessed through your Muse account. If you are familiar with using spreadsheets and are confident finding your own way around new technology, there is a comprehensive tutorial built into the software which is adequate for getting you started with the basics.

Data is entered into the SPSS Data Editor. This consists of two viewers: The Variable viewer which enables you to define and view the different fields or variables for the data you have collected. The Data viewer for entering and viewing the data for each of the variable. Each row in the data viewer represents a 'case', *i.e.* all the data from a single questionnaire. Data is analysed using the Analyse option from the menu bar above the Data Editor. A number of different analysis options can be chosen from a drop down menu. You are likely to most commonly use the Descriptive Statistics option. Data can be presented in graphical form using the Graphs option, again from the menu bar above the Data Editor. Graphs can be edited using the Chart Editor. The results of performing analyses and producing graphs create Outputs which can be saved.

Strengths

Data can easily be imported from Excel or.csv files.

Limitations: Imported.csv files may not contain variables in the appropriate numerical format for SPSS, particularly if these have been exported from another application such as MOLE. You may have to spend a little time correcting the variable definitions in the variable viewer. Text fields cannot accommodate more than 255 characters and many open question responses may exceed this limit. Data imported from large comments fields in MOLE surveys or Excel spreadsheets in which fields can accommodate more than 255 characters may therefore be lost in the transfer process and may need to be added manually, or collated using an alternative application such as Word.

Practicalities

Questionnaire data from the evaluation tool and tracking tool in MOLE can be exported as a.csv file which can be imported into SPSS.

For Likert scale responses, MOLE will export the text answer selected rather than the numerical value of the scale associated with it. For the standard multiple choice Student Evaluation Questionnaires (SEQ's) the Management School has developed an Excel spreadsheet into which data exported from MOLE can be copied and which includes macros to perform the appropriate data conversions and data processing and analysis. They have given permission for LeTS to adapt it to meet the needs of other departments. For initial advice about this tool contact the MOLE Support team in LeTS.

It may be necessary to create more than one text variable in SPSS to accommodate the data from a single open question or comments field on a questionnaire because of the 255 character limit on a single field.

5

Research Method

INTRODUCTION

There are many ways to get information. The most common research methods are: literature searches, talking with people, focus groups, personal interviews, telephone surveys, mail surveys, e-mail surveys, and internet surveys.

A *literature search* involves reviewing all readily available materials. These materials can include internal company information, relevant trade publications, newspapers, magazines, annual reports, company literature, on-line data bases, and any other published materials. It is a very inexpensive method of gathering information, although it often does not yield timely information. Literature searches over the web are the fastest, while library literature searches can take between one and eight weeks.

Talking with people is a good way to get information during the initial stages of a research project. It can be used to gather information that is not publicly available, or that is too new to be found in the literature. Examples might include meetings with prospects, customers, suppliers, and other types of business conversations at trade shows, seminars, and association meetings. Although often valuable, the information has questionable validity because it is highly subjective and might not be representative of the population.

A *focus group* is used as a preliminary research technique to explore peoples ideas and attitudes. It is often used to test new approaches (such as products or advertising), and to discover customer concerns. A group of 6 to 20 people meet in a conference-room-like setting with a trained moderator. The room usually contains a one-way mirror for viewing, including audio and video capabilities. The moderator leads the group's discussion and keeps the focus on the areas you want to explore. Focus groups can be conducted within a couple of weeks and cost between two and three thousand dollars. Their disadvantage is that the sample is small and may not be representative of the population in general.

Personal interviews are a way to get in-depth and comprehensive information. They involve one person interviewing another person for personal or detailed information. Personal interviews are very expensive because of the one-to-one nature of the interview ($50+ per interview). Typically, an interviewer will ask questions from a written questionnaire and record the answers verbatim. Sometimes, the questionnaire is simply a list of topics that the research wants to discuss with an industry expert. Personal interviews (because of their expense) are generally used only when subjects are not likely to respond to other survey methods.

Telephone surveys are the fastest method of gathering information from a relatively large sample (100-400 respondents). The interviewer follows a prepared script that is essentially the same as a written questionnaire. However, unlike a mail survey, the telephone survey allows the opportunity for some opinion probing. Telephone surveys generally last less than ten minutes. Typical costs are between four and six thousand dollars, and they can be completed in two to four weeks.

Mail surveys are a cost effective method of gathering information. They are ideal for large sample sizes, or when the sample comes from a wide geographic area. They cost a little less than telephone interviews, however, they take over twice as long to complete (eight to twelve weeks). Because there is no interviewer, there is no possibility of interviewer bias. The main disadvantage is the inability to probe respondents for more detailed information.

E-mail and internet surveys are relatively new and little is known about the effect of sampling bias in internet surveys.

While it is clearly the most cost effective and fastest method of distributing a survey, the demographic profile of the internet user does not represent the general population, although this is changing. Before doing an e-mail or internet survey, carefully consider the effect that this bias might have on the results.

THE LAYMAN'S GUIDE TO SOCIAL RESEARCH METHODS

INTRODUCTION

Understanding the basics of social research methods can, at times, feel as if one is walking a very fine line between complete comprehension and abysmal failure. The terminology is complex, the concepts highly interrelated, and as we all know, every little detail matters when it comes to a successful thesis or dissertation. As decided a sort of "cliff notes" to my experience in two good courses on this topic may be useful (if not amusing) to others who are in the same position as I am-a weary eyed student about to begin a dissertation. It is, as you know a daunting task filled with both anticipation and a lot of fear.

STEPS FOR A SUCCESSFUL POLICY ANALYSIS

WHAT IS POLICY ANALYSIS

"The process through which we identify and evaluate alternative policies or programs that are intended to lessen or resolve social, economic, or physical problems."

POLICY ANALYSIS IN SIX EASY STEPS

Based on the ideas and approach followed by Carl V. Patton there exists a very simple pattern of ideas and points to be considered in doing an actual policy analysis. The six steps are as follows:

Verify, define, and detail the Problem

The most relevant and important of them all because

many times the objectives are not clear or even contradictory from each other. A successful policy analysis will have allocated and identified clearly the problem to be resolved in the following steps. This is the foundation for an efficient and effective outcome of the whole process. The analyst must question both the interested parties involved as well as their agendas of the outcome. Locating the problem in such a way that eliminates any ambiguity for future references.

Establish Evaluation Criteria

In order to compare, measure and select among alternatives, relevant evaluation criteria must be established. In this step it must be considered cost, net benefit, effectiveness, efficiency, equity, administrative ease, legality, and political acceptability. Economic benefits must be considered in evaluating the policy. How the policy will harm or benefit a particular group or groups will depend on the number of option viable Options more difficult than others must be considered but ultimately decided through analyzing the parties involved with policy. Political and other variables go hand in hand with the evaluation criteria to be followed. Most of the time the client, or person or group, interested in the policy analysis will dictate the direction or evaluation criteria to follow.

Identify Alternative Policies

In order to reach this third step the other two must have been successfully reached and completed. As it can be seen, the policy analysis involves an incrementalist approach; reaching one step in order to go on to the next. In this third step understanding what is sought is very important. In order to generate alternatives, it becomes important to have a clear understanding of the problem and how to go about it. Possible alternatives include the "do nothing approach" (status quo), and any other that can benefit the outcome. Combining alternatives generates better solutions not thought of before. Relying on past experiences from other groups or policy analysis helps to create a more thorough analysis and understanding. It is important to avoid settling prematurely on a certain number of options in this step; many options must be considered before settling into a reduced number of alternatives. Brainstorming, research, experiments, writing scenarios, or *concept mapping* greatly help in finding new alternatives that will help reach an "optimal" solution.

Evaluate Alternative Policies

Packaging of alternatives into strategies is the next step in accomplishing a thorough policy analysis. It becomes necessary to evaluate how each possible alternative benefits the criteria previously established. Additional data needs to be collected in analyzing the different levels of influence: the economical, political and social dimensions of the problem. These dimensions are analysed through quantitative and qualitative analysis, that is the benefits and costs per alternative.

Political questions in attaining the goals are analysed as to see whether they satisfy the interested parties of the policy analysis. In doing this more concise analysis the problem may not exist as originally identified; the actual problem statement from the first step may suffer a transformation, which is explained after evaluating the alternatives in greater detail. New aspects of the problem may be found to be transient and even different from the original problem statement. This modification process allows this method of policy analysis to allow for a "recycling" of information in all the steps.

Several fast interactions through the policy analysis may well be more efficient and effective than a single detailed one. What this means is that the efficiency is greatly increased when several projects are analysed and evaluated rather than just one in great detail, allowing for a wider scope of possible solutions. Patton further suggests to avoid the tool box approach: attacking options with a favourite analysis method; its important to have a heterogeneous approach in analyzing the different possible alternatives.

It becomes inefficient to view each alternative under a single perspective; its clearly relevant the need to evaluate each alternative following diverse evaluating approaches singled out according to the uniqueness of each of them.

Display and Distinguish among Alternative Policies

The results of the evaluation of possible alternatives list the degree to which criteria are met in each of them. Numerical results don't speak for themselves but are of great help in reaching a satisfying solution in the decision. Comparison schemes used to summarize virtues are of great help in distinguishing among several options; scenarios with quantitative methods, qualitative analysis, and complex political considerations can be melded into general alternatives containing many more from the original ones. In making the comparison and distinction of each alternative it is necessary to play out the economic, political, legal, and administrative ramification of each option.

Political analysis is a major factor of decision of distinction among the choices; display the positive effects and negative effects interested in implementing the policy. This political approach will ultimately analyse how the number of participants will improve or diminish the implementation. It will also criticize on how the internal cooperation of the interested units or parties will play an important role in the outcome of the policy analysis. Mixing two or more alternatives is a very common and practiced approach in attaining a very reasonably justified policy analysis.

Monitoring the Implemented Policy

Assure continuity, determine whether they are having impact. "Even after a policy has been implemented, there may be some doubt whether the problem was resolved appropriately and even whether the selected policy is being implemented properly.

This concerns require that policies and programs be maintained and monitored during implementation to assure that they do not change for unintentionally, to measure the impact that they are having, to determine whether they are having the impact intended, and to decide whether they should be continued, modified or terminated." Mainly, we are talking about internal validity; whether our programs makes a difference, if there is no other alternate explanations. This step is very important because of the special characteristic that program evaluation and research design presents in this particular step.

RESEARCH METHOD

The scientific method is the means by which researchers are able to make conclusive statements about their studies with a minimum of bias. The interpretation of data, for example the result of a new drug study, can be laden with bias. The researcher often has a personal stakes in the results of his work.

SCIENTIFIC METHOD

The scientific method is the means by which researchers are able to make conclusive statements about their studies with a minimum of bias. The interpretation of data, for example the result of a new drug study, can be laden with bias. The researcher often has a personal stakes in the results of his work. As any skilled debater knows, just about any opinion can be justified and presented as fact. In order to minimize the influence of personal stakes and biased opinions, a standard method of testing a hypothesis is expected to be used by all members of the scientific community.

How does the Scientific Method Work

The first step to using the scientific method is to have some basis for conducting your research. This is based on observed phenomena that is either directly or indirectly related to the specific subject matter of your proposed research. For example, you may have observed that Drug A is effective in treating an illness(Disease A) caused my Virus A. A new illness (Disease B) has arisen that mimics some symptoms of Disease A, but with variation (such as the patient with Disease B has swollen lymph nodes and a low grade fever instead of no swelling and a high grade fever).

Outbreaks of Disease B occur near outbreaks of Disease A. These are the observations you make in your first step of using the Scientific Method.

The next step is to form a hypothesis to explain some aspect of your observations. You speculate that the virus that causes Disease B is either Virus A or it is related to Virus A. Your hypothesis is that the cause of Disease A and Disease B is the same virus. Now that you have a hypothesis, you are ready to test it. You must now use your hypothesis to predict other phenomena that have not yet been observed. You know that Drug A will wipe out Disease A. If Disease B is caused by the same virus, you reason that the same drug should be effective.

The final step of the scientific method is to rigorously test your prediction. Remember, you cannot "prove" your hypothesis. You can only fail to disprove it. While this is an example of how the scientific method is used in everyday research and hypothesis testing, it is also the basis of creating theories and laws. The scientific method requires a hypothesis to be eliminated if experiments repeatedly contradict predictions. No matter how great a hypothesis sounds, it is only as good as it's ability to consistently predict experimental results. It should also be noted that a theory or hypothesis is not meaningful if it is not quantitative and testable. If a theory does not allow for predictions and experimental research to confirm these predictions, than it is not a scientific theory.

WHAT IS A HYPOTHESIS

It is important to distinguish between a hypothesis, and a theory or law. Although in everyday language, people sometimes use these terms interchangeably, they have very distinct connotations in the scientific community.

A hypothesis is a 'small' cause and effect statement about a specific set of circumstances. It represents a belief that a researcher possesses before conducting a satisfactory number of experiments that could potentially disprove that belief. For example, you open your refrigerator at home and are greeted with a horrible sour smell. You decide that the milk must have gone bad. This is your hypothesis. It is based on the phenomena your are observing right now (sour smell) as well as knowledge from past experience (bad milk has a sour smell). You test your hypothesis by opening the container of milk and smelling it. You find that the milk doesn't smell sour after all, so you must come up with another hypothesis (maybe it is the leftover lasagna from last week!).

A theory or law in the world of science is a hypothesis, or many hypotheses, which have undergone rigorous tests and have never been disproved. There is no set number of tests or a set length of time in which a hypothesis can become a theory or a law. A hypothesis becomes a theory or law when it is the general consensus of the scientific community that it should be so. Theories and laws are not as easily discarded as hypotheses.

MISAPPLICATIONS OF THE SCIENTIFIC METHOD

A common error encountered by people who claim to use the scientific method is a lack of testing. A hypothesis brought about by common observations or common sense does not have scientific validity. As stated above, even though a good debater may be quite convincing as he conveys the merits of his theory, logical arguments are not an acceptable replacement for experimental testing.

Although the purpose of the scientific method is to eliminate researcher bias, an investigation of the raw data from an experiment is always a good idea. Researchers sometimes toss out data that does not support their hypothesis. This isn't necessarily done with the intent of deception, it is sometimes done because the researcher so passionately believes in his hypothesis that he assumes unsupportive data must have been obtained in error. Other times, outside forces (such as the corporation sponsoring and conducting the research) may put extreme pressure on the researcher to get specific results.

The best way for the scientific community, and the general public, to deal with these errors is to promote multiple, independent experiments. We are all familiar with "breaking news" (that seems to break nearly every day!) about a new miracle drug or herbal remedy. In most cases, this "breaking news" was released by a single source—usually a source with financial stakes in the new miracle. Look for multiple sources to confirm a hypothesis before you hand your money over for a new product. If possible, also try to discover where the funding came from in these experiments. You may have three different lab reports, all confirming that Drug A is the most effective cure, but if all three laboratories are funded by the same drug company-you may want to raise an eyebrow.

RESEARCH

Research is literally everywhere, and knowing about research methods will help us understand how we came to know what we accept as fact. We all know that 4 out of 5 dentists recommend sugarless gum for their patients who chew gum, and that taste-tests show that Burger King's Croissanwich is preferred 2 to 1 over McDonald's Egg McMuffin. We also know that proper nutrition is important for children's development, and that heroin is addictive. The question is: How did we come to know these things?

We learned about these things through research. Someone somewhere did a study and found each of the above findings. While most of us accept the value of good nutrition for children and the dangers of heroin, at least some of us (especially Certs Sugarless Mint and Egg McMuffin fans) would quibble with the other two findings. Just how did the researchers do the studies that found those numbers (by the way, get in the habit of questioning how researchers find their results, even those you agree with)?

Who knows, maybe they did the Croissanwich study at Burger King one morning and asked people which sandwich they preferred. The fact that the participants had already gone to Burger King for breakfast meant that they had at least some tolerance for Croissanwiches. What about the dentists? What percentage of them "recommended" candies of any kind? Which dentists participated in the study and what exactly do they prefer about sugarless gum over sugarless mints? When we ask questions like these, we are questioning the methods the researchers employed in their studies. Research methods are a variety of techniques that people use when studying a given phenomenon. They are planned, scientific, and value-neutral.

What that means is that good research methods don't "just happen." Instead, they are deliberately employed in a way that is designed to maximize the accuracy of the results. Assume, for example, that you want to do a survey to assess students' level of satisfaction with the parking situation on your campus. Because no one wants to take time to be interviewed, you are forced to start interviewing the people sitting in their cars waiting for parking places. Since they're bored anyway and have nothing better to do, they agree to answer your questions. Lo and behold, you find that all 50 people you talk to are VERY unhappy about the lack of parking spaces on campus.

Based on your research, can you say that the students at your university are dissatisfied with the parking situation? Even if the students in general really are unhappy, your study doesn't provide a strong answer to the question. What about the other 9,950 students that attend your school? What do they think? If they are all happy, then your 50 research participants represent less than one per cent of the students. Ooops, major problem! Your sample was biased. You unnecessarily increased your chances of error by relying on a bogus sample. A better way for you to do your survey would be to *randomly* select names from a list of all enrolled students (there will be more on random samples in a later section). You could ask them what they think about the parking situation. Because your sample was randomly selected, you would expect their answers to reflect what students as a whole felt about the parking situation. The use of random samples is just one way that researchers try to ensure that the answers they find are accurate.

Some common research techniques are surveys, experiments, and field research. No matter what technique you use, you will pay attention to certain things. The *sample,* or people you study, is one very important concern as we learned above. Studying the wrong group of people means that others can criticize your study. You often see this happen when magazines and other publications report the results of their monthly polls.

Who answers these polls, anyway? The question other researchers have is usually: How *representative* of the rest of the population is the sample? Just because 37 out of 50 women surveyed by National Enquirer preferred chocolate to sex doesn't mean anyone else does (this was a real survey reported on 10/11/94!).

Another way that researchers try to increase accuracy in their findings is by trying to ask questions in a non-biased, easy-to-understand manner. How a question is worded can affect the responses that people provide. Because people react to subconscious cues contained within questions, for example, lawyers are not allowed to ask witnesses "leading" questions in court (*i.e.,* questions that suggest an answer, such as "He was standing at the top of the stairs, wasn't he?").

SELECTIVE OBSERVATION

Selective observation happens when our attention is drawn to answers or observations that confirm our pre-existing beliefs. It's a lot like selective hearing (*i.e.,* when people, especially children, hear only the things they want to hear). For example, if I hypothesize that blacks are more likely than whites to speed, I am probably likely to note the blacks who are speeding while paying less attention to speeding whites and blacks who are not speeding. A better approach to this study would be to write down the speed of every car going by and the race of the driver. I could then make tables and compare the percentages of speeding drivers of each race.

A few ways you can try to avoid selective observation in your research are to do a literature review (so you'll know which relationships other researchers found), decide your research approach beforehand (*e.g.,* when I decided above to write down the speed of every car rather than just depend on my memory), take thorough notes (to prevent biases from affecting your memory), watch for "disconfirmatory" information (such as speeding whites and non-speeding blacks), and consider both "sides" of your study (*i.e.*, try to argue against your hypothesis; if you can't poke holes in your theory then you're more likely to be on the mark).

You could also use time or area sampling. Time or area sampling means that you focus your attention on a smaller part of the action for a given amount of time. Instead of trying to watch the entire crowd at a basketball game, for example, I could look at the rightmost four columns of people for ten minutes, then the next four columns for ten minutes, and so on. By doing this, I would be able to get a picture of what the whole crowd was doing. And, most importantly for selective observation, I would be forced to look at all parts of the crowd at some point time, rather than just those who were doing what I expected them to do.

INACCURATE OBSERVATION

Inaccurate observation happens when we "misremember" or misrecord data. How many times have you missed a question on an exam because you incorrectly copied down something from lecture? That's one form of inaccurate observation. You thought you correctly observed the information when you really hadn't done so. Have you ever misunderstood what someone said, and thought s/he said something that rhymed with his/her actual utterances? That's another form of inaccurate observation. Your brain somehow miscoded the information at the processing stage. Inability to remember what you saw is another form. Your brain somehow jumbles or changes the original memory during the recall stage.

If you plan to observe (and possibly even take notes), you can increase your accuracy. Planning includes focusing on the task at hand rather than daydreaming during data collection; this will help prevent many errors. Doing a literature review can help by alerting you to what you might find (that way you won't be as surprised and will be better able to process and recall the information you see or hear). Other techniques include using forms (*e.g.*, where you circle people's characteristics rather than trying to write them down; forms also "guide" your observations to include all the important information points), doing time/area sampling (to reduce the amount and variety of information you are responsible for writing down and remembering at any given moment), and writing down as much information as you can (you should assume that anything you don't write down will be forgotten or remembered incorrectly). Another suggestion is to practice observing and recording your observations before actually doing it for real. Practice will make it easier to quickly and accurately record or summarize a given situation, and will show you the parts of the task about which you are less clear.

OVERGENERALIZATION

Overgeneralization is *generalizing* to others who are different from one's research population. This happens all the time. A study in City finds a high proportion of gang members and school administrators in rural America panic and institute draconian measures intended to stem the proliferation of gangs in their districts. A parenting program works in one community, so planners automatically assume it will work in theirs.

A new study program raises grades for high school students, so colleges clamor to include it in their curriculum. All three of these examples illustrate how quickly some people take research findings as absolute. Who knows, maybe the findings actually do generalize. The problem is that you can't make that assumption.

Just because a program works in one community doesn't mean that it will work in another. For example, consider the NCI Alcohol Treatment program in Navajo country. Alcohol abuse programs "borrowed" from dominant society didn't have very high success rates for American Indians, but that didn't stop social service agencies from using them anyway. Then, NCI incorporated traditional practices and ideas (*e.g.*, use of certain ceremonies) into their program which works much better for American Indians.

Not ironically, non-Indians (and many Christianized American Indians) don't find the program to be very useful. This example illustrates overgeneralization because the social service agencies assumed that dominant society treatment programs would work equally well for American Indian clients.

After all, they worked for whites! The social service agencies were overgeneralizing; that is, they were assuming that what works for whites will work for others. In the end, they learned that what's good for the goose is not always good for the gander. There might be important differences between populations that will affect the success of a given program.

Some ways to avoid overgeneralizing:

- Replicate one's study (to ensure that the results apply to different populations),
- Support many tests of the same theory (to make sure it operates the same way under different circumstances and with varying populations),
- Attempt to use representative samples (to reduce the likelihood of aberrant findings due to the unique sample), and
- Recognize the limitations of one's research (possibly the most important; don't claim that your study is the definitive work unless you've paid very close attention to design and implementation).

MADE-UP INFORMATION

Made-up information happens when one fills in details without a scientific basis for doing so. As researchers, we have to fill in a lot details; this is called inferring. The problem is that some inferring isn't based in science; instead, it's based in stereotype and speculation.

In one of my research classes long ago, a classmate observed and reported to us the following: an attractive young woman was sitting alone in a bar for a few minutes when a man walked in. The two talked briefly, then left together smiling. What happened in that situation? At first, my classmate thought it was a typical "pick up" and may have even been the solicitation of a prostitute (especially since neither consumed any beverages). When he saw this happen day after day at the same time, however, he approached the young woman and asked her to explain what he had been observing. It turned out that she worked in the area, and merely came to sit in the bar to wait for her brother to pick her up after work.

One of my students observed a number of student government meetings and concluded at first that the council was unprofessional and didn't respect each other's views. She based this conclusion on the fact that throughout all the meetings, they were passing notes, especially during discussion of what she felt were important issues. It later turned out that the students were merely following Robert's Rules of Order, and were writing their names on slips of paper to be passed to the student who maintained the speaker's list. He would add the names from the slips of paper onto the bottom of his list and call on them in turn. Oops, major misunderstanding. Had she asked earlier, she would have saved herself a lot of embarrassment! Made-up information may or may not be correct.

For wonderfully humourous examples of this problem, consider Horace Miner's contention that the Nacirema are a "magic-ridden people" who are destined to self-destruct due to their mystical beliefs and an unknown scholar's equally clever discussion of the Asu's "overwhelming preoccupation with the care and feeding" of their sacred racs In case you haven't figured it out, the Nacirema and Asu are 'Americans' (both are simply spelled backwards to baffle the reader) and the writers are poking fun at anthropologists who misinterpret what they see.

By far the best example of made-up information is to be found in David Macaulay's Motel of the Mysteries, a futuristic look at an excavation of the Toot'n'C'mon Motel in the year 4022. Here's a sample of the misunderstandings that the future archaeologists made, this text describing a common credit card found in one of the rooms: This extremely fine piece of workmanship served as a portable shrine which was to be carried through life and into eternal life. Its delicate inscriptions were intended to identify an individual's religious preference along with the burial site to which the body should be delivered when necessary. Matching inscriptions were found on the main doors of the sanctuary. Because the ancients were unable to predict the exact time of death, each of the shrines had to last for an entire year.

Instead of making up information, you could do a variety of things. First, you could do like my classmate above and ask someone who knows what's going on for their opinion. If you're going to be in a strange or new setting, you could do a thorough literature review to help you understand possible outcomes and unique customs (this would have helped my student understand the note passing behaviour she observed). Most of all, you should rely on prior research studies and/or theory to guide your own interpretation of what was going on.

EX POST FACTO HYPOTHESIZING

Ex post facto hypothesizing happens when a researcher decides what happened after it happened and after the study was done. In scientific research, we have to decide what will happen before we do our study, not after. What this means is that our research must be guided by theory, and that we have made predictions (hypotheses) about what we will find. It's similar in some respects to Ex Post Facto laws (laws that have been enacted after the offensive act took place), and are just as repulsive, at least to researchers.

This doesn't mean that one cannot use data that has already been collected by others. Assume, for example, that you wish to research whether election outcomes are based on public perceptions about the economy. You could obtain from a data clearinghouse, election data from the past. Then, you could somehow assess public perceptions of the economy at the time of each election (through polls from the era, editorials in newspapers, or some other method). You could then correlate the two and test your theory. What differentiates your study from a common case of Ex Post Facto Hypothesizing is that you made your predictions, then tested them. You didn't form your hypothesis before-hand.

The ways to avoid this problem in your own research are simple. First, remember that correlation does NOT imply causation; in other words just because the birth rate is high in areas with lots of storks doesn't mean that the storks are bringing the babies. Then, don't feel obligated to report your research as deductive (that is based on hypothesis testing). If you were simply exploring a topic, own up to it. Exploratory research is very helpful in establishing foundations for future scholars.

ILLOGICAL REASONING

Illogical reasoning is just that: illogical. It ranges from a bit off the mark to absolutely absurd. For example, we have all heard that bread always lands buttered-side down, but seldom question where that knowledge came from. I've actually heard of someone who proposed (rather humourously) that if one took a piece of buttered bread and tied it to the back of a cat and dropped the resulting mass, that the two diametrically opposing forces would cause the bread and cat combo to ceaselessly spin several inches from the floor. It's illogical to assume that the butter side will always land down. I've actually done research that shows that bread (and bagels and crackers and other such things) will land buttered-side up as well (now you know what to do with that horrible dorm toast). Now, the cat is another story; research has repeatedly shown that cats somehow miraculously upright themselves from even rather short falls (just take my word for this; don't try it on the family feline).

Logically flawed assumptions in research. Medical researchers in 1879 argued that masturbation causes insanity and recommended exercise to the point of fatigue as a treatment. The same journal contains research that claims that insanity can be contagious if people live together. In other words, simply masturbating or marrying someone with a history of insanity could make you insane. Despite its lack of logic, this research does add new meaning to Rosenhan's research that showed that insanity was assumed for anyone who had been admitted to the 12 mental hospitals under study. As a side note, the journal also includes a writeup on coffee as a cure for typhoid fever. Pretty illogical, eh?

P.S. Just because you reason logically doesn't mean that you're right about what you saw. Think about all the times you've made wonderfully sound assumptions about outcomes and been wrong. We can't all be like Sherlock Holmes and be right every time.

Two ways to avoid illogical reasoning in your own research are to base your decisions on prior research and theory, and to make extensive use of peer review. Peer review is when you have others read and critique your writings; this helps scholars find places where they've reasoned illogically.

EGO INVOLVEMENT IN UNDERSTANDING

Ego involvement in understanding happens when people let the human side of them dictate their findings and how they view findings by other researchers. Those who are members of a researched group (*e.g.*, sorority, family, or religion), for example, may find it difficult to explain what happens without bias; they would feel obligated to protect the "deep, dark secrets" of the group. They may feel compelled to justify even sinister behaviour.

Ego involvement can also affect research on a person's "pet project." If I've just spent ten years developing a new treatment program, someone else should do the evaluation because I might be tempted to fudge results here and there, or at least misinterpret things in a way that is favourable to my program. After all, I worked hard on the program; it must really work! Those whose jobs depend on the program should also be excluded from the evaluation team because it would be hard to disband even an ineffective program that provided one's salary.

Sometimes ego involvement affects how we look at and react to research by others. Studies whose findings oppose what we believe are easily dismissed despite their rigor, whereas even shoddy studies that support our viewpoints are cited ad nauseam. Unfortunately, ego involvement by journal editors sometimes determines which studies get into print.

The best way to avoid ego involvement is to try to remain neutral, or to stay away from topics about which you can't be neutral. Be honest with yourself and refrain from research on topics that might be too close to home for you. Another suggestion is to use the team approach. It's harder to get too ego involved when sharing the research task with others who don't have your intense views on a given topic.

PREMATURE CLOSURE OF INQUIRY

Premature closure of inquiry occurs when we decide that we know enough about a topic and decide that it no longer warrants future study. Some topics are fairly easily dismissed as trite (*e.g.*, science is now certain the earth revolves around the sun). Other topics, however, are still misunderstood (*e.g.*, the causes of crime and insanity). For these less understood topics, it's important to keep researching until we more fully understand them.

Sometimes research is too controversial to get funding or support, regardless of its importance. Studies that parallel those by Stanley Milgram (who did the shocking experiments on obedience to authority) and Philip Zimbardo (whose Stanford prison experiment had to be called off early due to psychological trauma experienced by some of the participants) may have difficulty getting support in today's world. Likewise, medical research like that conducted by notorious Nazi Josef Mengele (whose fascination with flawless research methods led him to literally murder set after set of twins in the name of research) has been condemned by all civilized nations.

The best way to deal with premature closure is to keep looking for answers, even if it involves using different approaches. While the findings of Milgram's and Zimbardo's studies were valuable to society, the toll taken on their subjects was too great. Mengele's methods were simply satanic at best. Contemporary researchers shouldn't quit doing research on medical issues, obedience to authority, and socialization of prison inmates and guards; instead, they should find non-controversial ways to do it. This may require some serious thought and planning. Some contemporary researchers, for example, have studied obedience to authority by having people endlessly tear paper into little squares. Their willingness to comply with this task causes far less stress than the belief that they were administering sometimes fatal shocks to another human being while still allowing us to study the fascinating phenomenon.

MYSTIFICATION

Mystification happens when we attribute results to the supernatural. In olden days (and sometimes to this day), illnesses were blamed on a variety of spirits and deities. Trying to find a cure for cancer might have lead early researchers to see their local clergy! Crime, attributed to the devil's influence, was felt to result from lack of religious training.

Sometimes, the people we study will use mystification. One of my students, for example, asked ranchers about the mutilation of their livestock. To his surprise, a number of them insisted that the damage was caused by space aliens. Rather than report this as the actual cause, he kept looking for more sensible answers.

The ways to avoid mystification are similar to those mentioned above for other errors. First, keep looking for answers when you feel tempted to rely on enchanted explanations. Second, peer review will prevent you from embarrassing yourself when you insist that witches have afflicted young girls in Salem, Massachusetts.

SOME EARLY STEPS IN RESEARCH

CHOOSE A TOPIC

Choosing a one's topic is a very important step. If you have a choice of topics, you want to pick one that interests you and to which you have at least some access. Many researchers try to study topics that are fascinating, but cannot really be researched due to a variety of concerns. Sometimes travel expenses are too high, or the researcher cannot gain access. Patricia Adler's (1993) study of upper level drug dealers and smugglers, for example, could not be completed by just any researcher. Adler was somehow able to gain access to the dealers and their world. Likewise, researchers who want to know more about serial killers might have problems finding people to interview. Even "everyday" topics like why kids join gangs means the researcher may have to find some gang members to study. In other words, don't bite off more than you can chew.

The second problem is the breadth of the topic. You want to pick a topic that is "narrow" enough that your research is focused, but not so narrow that you can't find any information on the topic. A researcher who is interested in African-Americans in criminal justice, for example, might easily find him/herself overwhelmed by the breadth of the topic. If the topic is narrowed to African-American judges, there will be fewer problems, but the topic is still rather broad. A better choice would factors that affect the sentencing of misdemeanants by African-American judges. An example of a "too narrow" topic might be child custody decisions by lesbian judges. No doubt it's an interesting topic, but it probably doesn't have enough information for a student to do a research study on it.

Q1. Which of the following would be the best topic for an undergraduate student working on a one-term class project?

A. Teenagers' attitudes
B. Organized crime figure's beliefs about how to reduce poverty in inner city areas
C. The factors that students feel were important in their decisions to attend your university
D. How the *Aha Kuka O na Kupuna* courts of Hawai'i operate

REVIEW THE LITERATURE

This means reading other researchers' studies to learn how they did their research and what they found. It has been said that contemporary researchers are "great because they stand on the shoulders of those who have completed earlier research," that is, because they can utilize the knowledge learned by earlier researchers. Although it is very time consuming, reviewing the literature will help you decide which *variables* (*i.e.,* factors) are important to study, how to measure (*i.e.,* assess) them, and what findings to expect.

To illustrate the importance of "the literature," assume you have been assigned to do a study of the factors that influence the importance of K'e assumed by Naat'aniis during Peacemaking sessions. I'm sure that most people would want to first go to the library and find out what K'e, Naat'anii, and Peacemaking are. Even when I tell you that K'e can be considered to be "clan solidarity," that a Naat'anii is a respected elder or other person who is called upon to help resolve disputes, and that Peacemaking is a Navajo process during which disputants are brought together to work out their problems and restore themselves to "harmony," you will probably still be lost.

Reading about Navajo Peacemaking and studies on it will not only explain such confusing concepts, but will also point out to you variables that might be important in your study. After reading a number of studies, you might learn that "level of traditionality" is a major factor in the beliefs that a Naat'anii has about K'e, so that might be a good variable for you to include in your study.

You might also learn how to measure "level of traditionality," for example, through use of a pre-developed scale like the XYZ Traditionality Test. Knowing about the XYZ Traditionality Test could save you a lot of time because you won't have to design your own questions. Finally, you might also learn that other researchers have found that those people who are classified as more traditional are more likely to believe that K'e is important. You would expect to find the same thing in your study if you ask the same types of questions as were used in the earlier research. Remember, you can "borrow" anything you want from other studies as long as you give the authors proper credit. In fact, you are *expected* to use their work to make your study better. If you don't review the literature, you are missing out on a lot of good information and you will be considered "lazy" by other researchers.

DETERMINE THE RESEARCH QUESTION

The research question is the question you hope to answer with your study. The literature review will help you decide which questions are important, especially as you gain familiarity with the topic and begin to see its finer nuances. For example, you might discover while reading about college drop outs that students leave colleges they perceive as socially "distant" from them, and that minority students have higher attrition rates than academically comparable white students. Based on these two findings, you might decide to examine whether minority students are more likely to leave colleges they feel are unresponsive to their unique cultural needs.

Determining the research question is an important step because it will help guide the design of your research study. Assume, for example, that you are going to survey students who drop out of college. While you could selected from many research questions related to that topic, you have decided on: Are minority students more likely than whites to leave colleges that are unresponsive to cultural diversity issues? Now that you've determined what question you want to answer, you should make sure all your survey questions gather information related to the answers you seek. In other words, you may want to avoid asking questions related to the math requirements because that won't shed any light on your specific research question; you'll just be wasting your participants' time answering unrelated questions. All your questions must have a purpose, and that purpose is dictated by your research question.

You also want your question to have some level of interest. Try to avoid overly simple inquiries such as: Do students drop out of college? We already know the answer to that question (unfortunately, they do drop out). Just because a question has been asked and answered in the past doesn't prevent you from doing more research on it, however. You can ask new questions or otherwise extend the usefulness of earlier research.

Or, you can do a *replication* of an existing research study. Usually, replications are done to see if the findings from the first study *generalize* or apply to another time, place, or population. An example of a replication was done by an undergraduate student here at Rutgers University. She wanted to replicate Kenneth Clark's famous "doll" study. Back in the 1950's, Clark found that black children preferred to play with white dolls more than black dolls, and that they felt black dolls looked "bad" while white dolls were perceived as "nice." Clark argued that this was due to the black children's acceptance of negative stereotypes about their race. His findings played an important role in the eventual desegregation of public schools by the U.S. Supreme Court in Brown vs. Topeka Kansas Board of Education. More than 40 years later, a Rutgers-Camden student used Clark's approach and found that 4-7 year old black girls now prefer black dolls, except on one index item; they still feel the white dolls are "smarter" than the black ones. We could spend hours discussing what this replication shows us, but let us confine ourselves to acknowledging that replications can be interesting and valuable.

Q2. Which of the following questions would be least appropriate for a study that seeks to answer the research question: Are minority students more likely than whites to leave colleges that are unresponsive to cultural diversity issues?

A. Does XYZ University offer extramural sports programs?
B. Does XYZ University offer ethnic studies courses?
C. How satisfied are the white students with the cultural activities at XYZ University?

D. Do minority students at XYZ University feel the faculty don't respect their views?

DEVELOP A HYPOTHESIS

According to Vogt's Dictionary of Research Methods and Statistics (1993, Sage Publications), a hypothesis is: "A statement of the relationships among the variables that a researcher intends to study." This definition brings out two important ideas. First, hypotheses are assertions about how two or more variables are related to or associated with each another.

Assume that we're again interested in studying why students leave college. One possible hypothesis is: Students drop out of college because they cannot afford to pay the high costs. This hypothesis posits that dropping out and finances are related, and suggests that those who can't afford the costs will leave. It is testable, because we could see if students leave because they can't afford the cost (versus leaving for other reasons). Hypotheses should both say what we expect to find in our research and be testable.

The second concept the definition illustrates is the idea that hypotheses are formulated for variables that we *intend* to study. In other words, the hypotheses are formulated before we do our research. It's not considered proper to do the research, poke around in the data, and then write up a hypothesis that "fits" what we found. Hypotheses should be based on theory and what previous researchers have found; they are not just made up "off the cuff." Hypotheses are scientifically reasonable predictions.

Remember, all good hypotheses are testable, and make specific predictions about what the researcher expects to find in his/her research. The development and testing of hypotheses is one of the important things that makes research scientific.

Q4: Which of the following is NOT a good hypothesis?

A. Children from broken homes are more likely to join gangs.

B. Studying for exams using the SQ3R method will improve your grades
C. Reduced morals causes crime
D. Teenagers are more likely than older women to have low birth-weight babies

OPERATIONALIZATION

Operationalization is how we measure our concepts. Concepts are seldom easy to measure. Let's say I want to study how satisfaction with life affects school performance in college seniors. How would I recognize high "satisfaction with life" when I saw it? How could I make sure that other people would agree with me when I said that a certain person was or was not satisfied with his/her life? One way would be to read the literature and see what other people did.

Chances are that someone else out there has studied satisfaction with life and might have written up how to measure it. I could use their measures or make my own based on their ideas.

In fact, the items below were taken from one such scale, and you're welcome to use them as long as you provide a citation to the original creator of the scale:

- I couldn't be much happier.
- I get a lot of fun out of life.
- There is much purpose to what I am doing at present.

The next question is how to measure "school performance." Do I use grades, teacher recommendations, level of extra-curricular activities, or something else? You can see that operationalization requires a lot of thinking. We'll have a whole section on this later.

Let's pick something easier: What's a college drop out? Think about it for a minute or two. Drop outs are students who don't re-enroll in school, right? What if they leave for only one semester, then come back? Are they still drop outs? Is a student who leaves a college to go to Yale University still considered a drop out at the first college? Is someone a drop out who leaves for employment reasons, like those who leave because their recent promotion to vice president of their company leaves them little time this semester? What about women who leave school to have a baby, then return? You see that even "easy" ideas can be difficult to measure.

VARIABLES AND RELATIONSHIPS

Vriables and relationships! This is where a lot of students get confused. I'll admit that I found the differences tricky at first, but now they're both as natural as riding a bicycle.

The first thing we need to cover are the three main types of variables. The first type are *independent* variables (called IV's for short). These are predictor variables, or those that we expect to affect other variables.

The second type are dependent variables (called DV's for short). These are outcome variables, or those that we expect to be affected by other variables. IV's and DV's can be summarized pictorially as: *IV →DV* To help you, they can also be imagined as: *cause → effect*

The third type of variables are mediating variables. Mediating variables are those that we expect to affect some variables and to be affected by other variables. They can be summarized pictorially as:

IV → mediating variable → DV

Here's two examples to clarify the types of variables:

If I think that girls are more likely than boys to become secretaries, the IV is gender and the DV is type of career, as below:

Gender → Type of Career

If I feel that girls are more likely to be socialized into "feminine" careers (while boys are socialized into more "masculine" careers) and that makes them more likely to become secretaries, the IV is gender, the mediating variable is type of socialization, and the DV is type of career.

Gender → Level of Socialization → Type of Career

Q1: I hypothesize that smoking causes cancer. Which of the following variable types does "smoking behaviour" fit under?

A. IV

B. DV

C. Moderating variable

Now, you may have noticed that while I felt that girls would become secretaries, my variables weren't "girl" and "secretary." Instead, they were *gender* and *career type*. This is because "girl" isn't a variable, it's an attribute. Variables must be able to vary, hence their name. Variables are made up of attributes, so that a given child's gender could be either the attribute "girl" or the attribute "boy."

Likewise, the variable *career type* could be secretary, truck driver, doctor, dentist, construction worker, or a host of other attributes. When we want to talk about all those occupations using one word, we call them *career type*.

Q2: Which of the following is NOT a variable?

A. Size of corporation

B. Level of satisfaction with supervisor

C. Administrator

D. Number of years at the company

Now, on to the final part, relationships. When we say we're "in a relationship with someone," we're saying that we are somehow associated with that person. Sometimes, others can tell something about that person based on the fact that they are in a relationship with us; it would be hard, for example, to imagine a staunch conservative marrying a devoted liberal. Knowing that a person is a staunch conservative helps us picture their mate.

It works sort of the same way with variables. In research, two variables are related if knowing the value (or attribute) assigned to one helps us predict the value of the other. Assume for a moment those who studied at least five hours for the midterm exam in this class tended to get better grades, as in the table below:

	IV: number of hours studied for the midterm exam	
	< 5	**5 or more**
DV: grade C or less	9	1
B or better	1	9

Now, what grade would you predict that your classmate Joe received? The first thing you would want to know is how many hours Joe studied. If you learn that Joe studied for seven hours, you would guess he received a 'B' or better. The reason for this choice would be that 9 of the 10 people who studied for five or more hours got a 'B' or better. That's what bettors call darn good odds. Even if Joe is the one person who got a 'C' or less, you had a 90% chance of guessing his grade based on knowing just one factor about Joe (the number of hours he studied). We could easily say that number of hours studied and midterm grade are related. Of course, most relationships in the social sciences aren't quite so strong, which makes accurate predictions a bit harder to do. When knowing the value of one variable doesn't help us predict the other at all, we say the variables aren't related. To illustrate this, try to predict the grade received by a male classmate, using the table below.

		IV: gender of student	
		Male	**Female**
DV:grade	C or less	5	5
	B or better	5	5

There is no relationship here. Gender does not appear to be related to midterm grade; males and females are equally likely to get a 'B' or better. One way to look at relationships is to consider it as a betting game. While you might be willing to put money on a bet regarding a student who studied for seven hours, you would probably not want to put a whole lot of money on a student about whom all you knew was his/her gender. It would be too risky of a bet. Knowing gender doesn't help you predict grade enough to bet any money on the outcome. Whenever you would be willing to bet money on the outcome, the variables are related.

Q3: Which of the following scenarios best illustrates two variables that are related?

A. Gender and major, where 50% of sociology majors are male

B. GPA and career choice, where 1/2 of passing students become program administrators
C. Major and satisfaction with career, where 75% of sociology majors are satisfied with their career
D. Marital status and decision to go to college, where half of married people go to college

PORTRAIT OF THE YOUNG CHILD AS RESEARCHER

As all children, I have always relied heavily on research methods in my life. Even as young child, I participated in and mastered many techniques and theories of inquiry. This brief essay describes some of my experiences in the wonderful world of research.

My parents played a strong role in my development as a researcher. As a toddler, for example, I remember Mother baking holiday cookies. The delightfully sweet scent filled the kitchen and spilled over into the playroom in which I played. I was immediately drawn to the kitchen for a closer observation of the baking process (qualitative research is sometimes preferable to more quantitative approaches, especially when the concept under study defies quantification). I quickly discerned that the aroma emanated from the big white box in the corner with the window on the front door, and proceeded to conduct more thorough research when Mother yelled "HOT!" and slapped my hand as it moved towards the oven. As soon as she looked away, however, I firmly planted my hand on the glass window and shrieked in pain. I immediately recognized that we had interrater reliability, that is we both agreed that the oven was hot. There was no question in my mind about the validity of her initial assessment; my burnt fingertips were proof of that.

Later that day, Father came home and initiated his favourite game, Whasinabox. Whasinabox was designed to developdeductive reasoning skills. He would wrap objects in brightly coloured paper, sometimes enclosing them in boxes before applying the paper. The idea of the game was to ascertain the contents without first opening the package. He would often deliberately camouflage the contents to make it more difficult to determine what was in the box. We played this game every winter; he would go so far as putting the game pieces under an elaborately decorated tree to intensify the experience. I quickly learned a plethora of deductive skills, such as squeezing and shaking the packages to help me figure out what they contained.

Clothing, for example, didn't make rattling noises like many toys or board games did. I also learned from what was included in other boxes of the same general shape and size. My sister, on the other hand, preferred a more inductiveapproach. She would carefully cut the paper, look inside, and then re-tape the paper so that it looked as pristine as when Father put it under the tree. When it came time to announce what we hypothesized was inside the package, she would always make accurate "predictions." When Father discovered her techniques, however, he angrily accused her of ex post facto hypothesizing, that is coming up with one's hypothesis after a thorough examination of one's data (the contents of the package in this case). That her "prediction" was correct did not mitigate the fact that she hadn't played the game properly by using her deductive skills and facts available outside the package in formulating her hypothesis.

The winter ceremony was also a good way to learn a bit about sampling. Father would allow us to select which one of the many boxes under the tree we would open on Christmas Eve. This was an important decision because all of the other boxes had to wait until Christmas day. I usually engaged in simple random sampling. I would arrange the boxes on the floor then use a table of random numbers to decide which I would open first.

I thought this method was the best because each box had an equal chance of being selected. My sister, on the other hand, preferred to use a non-random sampling technique, purposive sampling. She always selected the largest box under the tree because it met her single criterion: big in size. To frustrate her, Father sometimes put a single piece of candy in the largest box, so that my selection method more often yielded a good sample. I told my sister if she would randomly select a box like I did, the contents of it would be morerepresentative of the other boxes, but she insisted on her haphazard method.

Another fond research memory centered on the value of operationalization in research. I learned through experience that my parents' opinions of things often differed vastly from my own, that is that we operationalized concepts differently. Father's ideas of what constituted good ways to spend family time, for example, never included rolling in the grass and playing in the mud.

Instead, he preferred more mundane activities, including attending church and long family "discussions" around the dinner table. Whenever Father would tell us how much we children would enjoy something, my instant reply would always be "and just how do you operationalize that concept, Father?" If I was lucky, I received an answer rather than a firm swat on the behind (and ironically, he always told me that the swats hurt him more than they did me leading me to wonder how he operationalized pain!).

My sister and I often engaged in rigorous hypothesis testing. We would take commonly accepted statements and attempt to test them through empirical research. One of these statements was that eating candy would make your teeth rot out. I told Mother how important it was to do research on the topic, and asked her to buy us several cases of candy; I and my sister would volunteer as subjects in this groundbreaking study.

She resisted, stating that research by others had firmly established a relationship between eating candy and tooth decay. My sister and I lamented that the external validity had not been confirmed; what if the results did not apply to other samples, areas, settings, and times? To address this possibility, we needed to conduct additional replications of the experiment in a variety of settings and with a variety of research subjects. By refusing to allow us to conduct this study, she was engaging in premature closure of inquiry. She would not yield, however, so I went to a backup funding source, Grandmother. Grandmother was much more agreeable and agreed to provide us with the required materials so that the valuable research could take place.

Now, it's your turn! Think about and write up a childhood experience in which you (or some other child) participated in and mastered some of the techniques and theories of inquiry. Send them to me, and I'll post them here.

5

Nano Research Report

LAYOUT OF RESEARCH REPORT

TOPICS AND REASONS

- Choose a current issue that is of interest to you and that is addressed in your partnership school
- How exactly does this current issue connect with teaching and learning?
- Briefly outline why you have chosen this issue. Why do you think it is an important issue to research?
- What kinds of questions, connected to your issue, would you like to find answers to?

Review of Literature and Research (400words)

- You need to read 3-4 articles in academic journals or books as well as any other material related to your issue or question
- Think about other resources that you can use such as the internet, reference books, magazines, libraries, CD-ROMS, videos, kits, pamphlets, handouts on particular programs that might be tied to the issue you are investigating
- What does the literature say about the issue?
- You should try to use direct quotes in your report and link this to the arguments you are attempting to present. Try not to quote big chunks from the articles you read. Rather attempt to paraphrase the intellectual ideas you are presented with. You must reference the source you are paraphrasing from

- Make sure you keep a record of all the bibliographical details of your references, as you must include a bibliography as part of the research. Failure to do so will end in a fail grade.
- Ensure that you make clear connections between the your writing an the literature you refer to!
- Are there areas in the literature that shed light into the questions you raise? Alternatively does any of the literature raise concerns for you? Why?

Data collection (250 words)

- You will need to interview at least one teacher at your partnership school.
- Ensure that you have planned out your questions before you interview your participant.
- Ensure that your interview questions are 'opened questions' rather than 'closed questions'
- Ensure that your questions directly relate to the issue you are concerned with
- You should also take field notes while in your partnership school and use them when writing up the report as part of your data
- You should also be writing a journal throughout your teaching experience. Your journal entries can be used as part of your data
- Will you use a survey?
- You need to outline why you choose particular types of methodologies for collecting data
- Think about how you will record the information that you gather. Will you take extensive notes?
- Ensure that you organise your material appropriately so that your task of analysis will be easier to take on

Data Analysis of the Data Gathered (300 words)

- Use subheadings
- Ensure that you make use of facts rather than opinions when incorporating the data into your writing and arguments

- Examine how teachers, administrators and students have responded to the issue you are investigating. How is this response tied to teaching and learning?
- Look for patterns of words or phrases in the data. This often is a clue as to the main messages that the participant wants you to take in
- Ensure that when you are analysing your data that you relate it back to the original question you seek to investigate at the beginning. Always tie your writing to the central focus questions you raise at the start of the report!

Discussion (450 words)

- What I NOW think about the question/issue- drawing together the ideas you have, the other research and your research to discuss your own thinking and knowledge and to identify any conclusions, recommendations, new questions that you think are important to investigate
- Is there anything that you could have done differently in your research design? If so what?

PREPARING DATA

Once all of the participants have completed the study measures and all of the data has been collected, the researcher must prepare the data to be analysed. Organizing the data correctly can save a lot of time and prevent mistakes. Most researchers choose to use a database or statistical analysis program that they can format to fit their needs in order to organize their data effectively.

A good researcher enters all of the data in the same format and in the same database, as doing otherwise might lead to confusion and difficulty with the statistical analysis later on. Once the data has been entered, it is crucial that the researcher check the data for accuracy. This can be accomplished by spot-checking a random assortment of participant data groups, but this method is not as effective as re-entering the data a second time and searching for discrepancies. This method is particularly easy to do when using numerical data because the researcher can simply use the database program to sum the columns of the spreadsheet and then look for differences in the totals. Perhaps the best method of accuracy checking is to use a specialized computer program that cross-checks double-entered data for discrepancies (as this method is free from error), though these programs can be hard to come by and may require extra training to use correctly.

DESCRIPTIVE STATISTICS

Descriptive statistics describe the data. They do not draw conclusions about the data. Descriptive statistics are normally applied to a single variable at a time. They can tell the researcher the central tendency of the variable, meaning the average score of a participant on a given study measure. The researcher can also determine the distribution of scores on a given study measure, or the range in which scores appear. Finally, descriptive statistics can be used to tell the researcher the frequency with which certain responses or scores arise on a given study measure. For example, in our imaginary study about the effectiveness of corrective lenses on economic productivity, the researcher might observe that the average dollars-per-week of a person with corrected vision is $500, whereas the average DPW for a person without corrected vision is $450.

A good researcher will know that this is not enough information to conclude that vision correction has an effect on economic productivity. Inferential statistics are necessary to draw conclusions of this kind. Descriptive statistics might also tell the researcher that the distribution of DPW is $351-$640 for the whole sample, and that the average DPW is $445 for the sample.

CORRELATION

Correlation is one of the most often used (and most often *mis*used) kinds of descriptive statistics. It is perhaps best described as "a single number that describes the degree of relationship between two variables." If two variables tend to be "correlated," that means that a participant's score on one tends to vary with a score on the other. For example, people's height and shoe size tend to be positively correlated. This means that for the most part, if a given man is tall, he is likely to have a large shoe size. If short, he is likely to have a smaller shoe size. Correlation can also be negative. For example, the temperature outside in Fahrenheit may be negatively correlated with the number of hot chocolates sold at a local coffee shop.

This is to say that as the temperature goes down, hot chocolate sales tend to go up. Although causality may seem to be implied in this situation, it is important to note that on a statistical level, correlation does not imply causation. A good researcher knows that there is no way to assess from*correlation alone* that a causal relationship exists between two variables.

INFERENTIAL STATISTICS

Inferential statistics allow the researcher to begin making inferences about the hypothesis on the basis of the data collected. This means that, while applying inferential statistics to data, the researcher is coming to conclusions about the population at large. Inferential statistics seek to generalize beyond the data in the study to find patterns that ostensibly exist in the target population. This course will not address the specific types of inferential statistics available to the researcher, but a succinct and very useful summary of them, complete with step-by-step examples and helpful descriptions, is available .

STATISTICAL SIGNIFICANCE

Researchers cannot simply conclude that there is a difference between two groups in a well-constructed study. This difference must be due to the manipulation of the independent variable. No matter how well a researcher designs the study, there always exists a degree of *error* in the results. This error can be due to individual differences both within and between experimental groups, or the error can be due to systematic differences within the researcher's sample. Irrespective of its source, this error acts as a kind of "noise" in the data.

It affects participants' scores on study measures even though it is not the variable of interest. Statistical significance is aimed at determining the probability that the observed result of a study was due to the influence of the independent variable rather than by chance. A result is "statistically significant" at a certain level. For example, a result might be significant at $p<.05$.

"P" represents the probability that the result was due to chance, and.05 represents a 5% probability that the result was due to chance. Therefore, $p<.05$ means that inferential statistical analysis has indicated that the observed results have over a 95% probability of being due to the influence of the independent variable. The 5% cutoff is generally thought of as the standard for most scientific research. Note that it is theoretically impossible to ever be entirely certain that one's results are not due to chance, as the nature of science is one of falsification, not immutable proof.

TYPES OF RESEARCH REPORT

Not all research that involves living individuals as subjects requires review by the human subjects committee. In general, research that may present even a slight element of risk must be reviewed, either by "expedited review" (review outside of the committee's normal schedule by committee staff) or by the full committee. For these purposes the concept of "risk" is broad indeed and includes not only obvious physical risk but also less-apparent risks such as those relating to confidentiality of data, possible stress to subjects, or possible damage to a subject's reputation or personal relationships. The following list indicates some types of research that may or may not require review.

Final determination of whether research is exempt should be made by the investigator—in consultation with a faculty or department adviser for student or staff investigators—only in unambiguous cases. For example, interviews with political candidates about their views on issues of public relevance, or anonymous surveys of adults on non-sensitive topics. If there is any doubt about whether review is required the investigator should consult the Chair of the IRB (631-244-5012 or PilottiM@dowling.edu) to discuss the project.

The mere fact that a research project requires review does not mean that there will likely be a problem, so long as the project is thoughtfully designed and proper consideration is given to the rights and welfare of the research subjects. Conversely, proposing a project that is exempt from the requirements of committee review does not relieve the investigator of any responsibilities relating to the research subjects; equal care must still be taken to ensure that subjects experience no harm to themselves or to their legitimate interests. If a research project includes some elements that may be exempt and some that may not, the more conservative judgment will obtain and the project should be submitted to the committee for review.

Review of Existing Records

Data already collected by another investigator or agency for research or other purposes, which are to be studied again or reanalysed. Identifiers may or may not be associated. It is helpful to know under what circumstances the data were originally collected—that is, whether subjects understood the future uses to which identifiable information about them might be put.

GUIDELINES

Research experience is as close to a professional problem-solving activity as anything in the curriculum. It provides exposure to research methods and an opportunity to work closely with a faculty advisor, graduate students, and sometimes post doctoral fellows and visiting scientists. Research usually requires the use of advanced concepts, a variety of experimental techniques, and state-of-the-art instrumentation. Ideally, undergraduate research should focus on a well-defined project that stands a reasonable chance of completion in the time available. A literature survey alone is not a satisfactory research project. Neither is repetition of established procedures.

Research is genuine exploration of the unknown that leads to new knowledge that often warrants publication. But whether or not the results of a research project are publishable, the project should be communicated in the form of a research report written by the student. It is important to realise that science depends on precise transmission of facts and ideas. Preparation of a comprehensive written research report is an essential part of a valid research experience, and the student should be aware of this requirement at the outset of the project. Interim reports may also be required, usually at the termination of the quarter or semester. Sufficient time should be allowed for satisfactory completion of reports, taking into account that initial drafts should be critiqued by the faculty advisor and corrected by the student at each stage.

Guidelines on how to prepare a professional-style research report are not routinely available. For this reason, the following information on report writing and format is provided to be helpful to undergraduate researchers and to faculty advisors.

Organization of the Research Report

Most scientific research reports, irrespective of the field, parallel the method of scientific reasoning. That is: the problem is defined, a hypothesis is created, experiments are devised to test the hypothesis, experiments are conducted, and conclusions are drawn. This framework is consistent with the following

Title and Title Page

The title should reflect the content and emphasis of the project described in the report. It should be as short as possible and include essential key words.

The author's name (*e.g.*, Mary B. Chung) should follow the title on a separate line, followed by the author's affiliation (*e.g.*, Department of Chemistry and Biochemistry, University of South Carolina, Columbia, SC 29208), the date, and possibly the origin of the report (*e.g.*, In partial fulfillment of a Senior Thesis Project under the supervision of Professor Suzanne G. White, June, 2006).

All of the above could appear on a single cover page. Acknowledgments and a table of contents can be added as preface pages if desired. The abstract should, in the briefest terms possible, describe the topic, the scope, the principal findings, and the conclusions. It should be written last to reflect accurately the content of the report. The lengths of abstracts vary, but seldom exceed 200-300 words.

A primary objective of an abstract is to communicate to the reader the essence of the paper. The reader will then be the judge of whether to read the full report or not. Were the report to appear in the primary literature, the abstract would serve as a key source of indexing terms and key words to be used in information retrieval. Author abstracts are often published verbatim in *Chemical Abstracts*.

Introduction

"A good introduction is a clear statement of the problem or project and why you are studying it.".

The nature of the problem and why it is of interest should be conveyed in the opening paragraphs. This section should describe clearly but briefly the background information on the problem, what has been done before (with proper literature citations), and the objectives of the current project. A clear relationship between the current project and the scope and limitations of earlier work should be made so that the reasons for the project and the approach used will be understood.

Experimental Details or Theoretical Analysis

This section should describe what was actually done. It is a succinct exposition of the laboratory notebook, describing procedures, techniques, instrumentation, special precautions, and so on. It should be sufficiently detailed that other experienced researchers would be able to repeat the work and obtain comparable results.

In theoretical reports, this section would include sufficient theoretical or mathematical analysis to enable derivations and numerical results to be checked. Computer programs from the public domain should be cited. New computer programs should be described in outline form.

If the experimental section is lengthy and detailed, as in synthetic work, it can be placed at the end of the report or as an appendix so that it does not interrupt the conceptual flow of the report. Its placement will depend on the nature of the project and the discretion of the writer.

Results

In this section, relevant data, observations, and findings are summarized. Tabulation of data, equations, charts, and figures can be used effectively to present results clearly and concisely. Schemes to show reaction sequences may be used here or elsewhere in the report.

Discussion

The crux of the report is the analysis and interpretation of the results. What do the results mean? How do they relate to the objectives of the project? To what extent have they resolved the problem? Because the "Results" and "Discussion" sections are interrelated, they can often be combined as one section.

Conclusions and Summary

A separate section outlining the main conclusions of the project is appropriate if conclusions have not already been stated in the 'Discussion' section. Directions for future work are also suitably expressed here.

A lengthy report, or one in which the findings are complex, usually benefits from a paragraph summarizing the main features of the report¾the objectives, the findings, and the conclusions. The last paragraph of text in manuscripts prepared for publication is customarily dedicated to acknowledgments. However, there is no rule about this, and research reports or senior theses frequently place acknowledgments following the title page.

Citing References

Literature references should be collated at the end of the report and cited in one of the formats described in The ACS Style Guide or standard journals. Do not mix formats. All references should be checked against the original literature. Never cite a reference that you have not read yourself. Double check all journal year, volume, issue, and inclusive page numbers to insure the accuracy of your citation.

Preparing the Manuscript

The personal computer and word processing have made manuscript preparation and revision a great deal easier than it used to be. Students should have the opportunity to use a word processor and have access to graphics software that allows numerical data to be graphed, chemical structures to be drawn, and mathematical equations to be represented. These are essential tools of the technical writer. All manuscripts should routinely be checked for spelling (programs to check spelling are helpful), and all manuscripts should be carefully proofread before being submitted. Preliminary drafts should be edited by the faculty advisor before the report is presented in final form.

BIBLIOGRAPHY AND ANNEXURE

Evaluation research: an introduction to principles, methods and practice. introduces fundamental principles of evaluation by demonstrating how a wide variety of social research methods can be applied in different evaluation contexts in imaginative and creative ways. Chapters include: paradigmatic choices in evaluation methodology; methods of data collection; evaluating criminal justice and crime prevention programmes, evaluation utilization.

This is a comprehensive introduction to research methodology, designed for self-instruction. The first part considers what research is, and the tools of research. Part 2 addresses focusing your research efforts, including stating the research problem, reviewing the literature, planning the research design and writing the research proposal. The next two parts look at qualitative and quantitative research methodologies, respectively. Finally, it advises on preparing the research report.

The book begins by examining the contemporary state of surveys within society and social science methodology, explaining the potential of the survey method and the ways it can be used effectively when resources are limited. It then takes the reader systematically through the process of conducting survey research, covering: the role of theory, the planning and design of projects, pilot work, access to informants, ethical issues, sampling methods, the preparation of questionnaires, interviewing, the use of computer packages, processing responses, statistical methods of data analysis, and the presentation of findings.

One chapter concerns secondary research, outlining the kinds of resources available. The main part of the book is about Internet-mediated research (IMR), in which the Internet is used to recruit participants, administer surveys and collect responses. It discusses the Internet as a viable research tool, whether it produces valid population samples, and various sampling methodologies. It reviews the technological issues and equipment relevant to IMR. Turning to the design and implementation of an Internet survey, it discusses design issues, and surveys by e-mail and on the Web (using HTML); and it considers problems that may need to be addressed, of equipment, methodology, netiquette and hackers. Finally, it presents three case studies to show how IMR can be implemented. This is a comprehensive text on social science research. Part 1 discusses background issues about research, including the meanings of methodology, the literature review, and ethical issues. Part 2 concerns planning and preparation, and the differences between qualitative and quantitative research designs. Part 3 covers quantitative data collection (experimental research, survey research, nonreactive research and secondary analysis). Part 4 covers qualitative data collection and analysis (field research and historical-comparative research). Part 5 advises on writing the research report; it also discusses the politics of social research. Each chapter starts with learning outcomes; contains illustrative examples, discussion questions and activities; and ends with a summary, questions for reflection and a list of further reading. The book progresses through the various stages of research: planning, reviewing existing research, choice of methodology, collecting organisational evidence, collecting and analysing qualitative data, using quantitative data, writing up the research, and reflecting on the relationship between HR research and HR practice.

This guide for first-time researchers is in two parts. Part 1 follows a step-by-step guide to the research process: developing objectives, designing the study, obtaining financial support, managing the research, writing the report and disseminating results. Part 2 is an introduction to some of the more common research methods involved in collecting and analysing quantitative and qualitative data and in doing desk research.

Department of Research Management

Funding

Information regarding funding opportunities for master's, doctoral and postdoctoral research is available at the Department of Research Management. The Department also administers funding from donor agencies such as the NRF, MRC, VODACOM, and others.

Conduct of the Research

The Department can advise and assist postgraduate staff and candidates with proposal and report writing, writing articles, aspects of research methodology, research budgets, and project planning. It should be noted that training in research skills and research methodology is usually provided by academic departments and faculties. The services provided by the Department of Research Management will be an add-on service, and will not replace what is offered by academic departments and faculties.

Research Development

Workshops are organised during the year for selected staff and candidates who are grant holders of specific development programmes. If space is available, others may also participate in such events.

Assistance with Statistical Analyses

The Department of Statistics at the NMMU provides a statistical support service to postgraduate candidates and staff members.

This service includes:

- One on one consultations between the candidate and a competent statistician on how the research needs to be structured in order to obtain valid data.
- An analysis of the data.
- Advice on how to interpret the results.
- Advice, if required, on relevant qualitative research techniques.

The statistical support service does not include:

- Compilation of documents (questionnaires, forms etc).
- Data coding.
- Data capturing.
- Ensuring the accuracy of the source data.

For more information on statistical support services contact the Department of Research Management.

NMMU Writing Centre

The Writing Centre offers guidance regarding the writing of all academic texts. Its services are aimed at helping candidates clarify their thought processes so that they can write more clearly. The consultants introduce candidates to argument development, cohesion and coherence, selection and integration of information sources, using appropriate academic register and referencing techniques, amongst others. Although editing is not part of their work, they can provide candidates with a list of editors who offer this service at a fee. See below for contact details.

NMMU Library Services

Information Retrieval

The Library offers access to many methods of retrieving information. Most of these are of an electronic nature, but the paper sources remain important. Information may be retrieved from:

- The Library Catalogue (OPAC)-OPAC is the computerized catalogue of the Library and is accessible on campus as well as off campus (by registering for off campus use through the Computing Centre);
- Reference works;
- Databases on CD ROM- CD-ROM journal indexes to various disciplines are available for use by registered members of NMMU Library;
- Online databases – Postgraduate candidates and staff can register with one of the information librarians in order to arrange for access to online databases;
- Periodical indexes and abstracts on CD-ROM and/or paper;
- Government Documents; and
- The Internet through the use of purchased units.

Information Services

Information Librarians assist with research requests, which should be filled out in detail on a Request for Information Form and submitted to one of the librarians.

Inter-lending services

The Inter-lending Department offers a service which enables users to have access to the collections of other libraries (nationally and internationally), subject to the Copyright Act.

Requests may be made on line via the candidate portal or OPAC. Inter-lending transactions are subject to differentiated costs.

Training

Postgraduate students and staff can make bookings for training relating to various aspects of information retrieval, including training in the use of databases and the Internet. The contact details for various sections of the library are provided below.

ICT Services

Access to Computing Facilities

Postgraduate candidates can register at ICT Services to obtain access to various computer laboratories for general use on South Campus (general purpose computer labs for other campuses in planning stage). Candidates will be provided with a user code, personalized e-mail address and can also open an account to obtain access to the Internet. See below for the contact details of ICT Services.

Training

The Department for Computer Science and Information Systems offer various computer literacy short courses at a cost to staff and candidates. Postgraduate candidates can contact the Department to obtain more information

Submission of final dissertation/thesis

ICT Services can also assist candidates in converting their final dissertations/theses for submission from Word (on condition that the document is in the required edited format) to PDF and to write the PDF version to CD Rom for submission to the Examinations Office.

On-campus accommodation available to postgraduate candidates at the NMMU

Postgraduate research candidates may apply for

accommodation in the Postgraduate Village or the Guest Quarters (also known as Student Village 7), which are designed specifically for housing postgraduate candidates, on the South Campus of the NMMU. The Postgraduate Village can accommodate 120 postgraduate candidates in one-, two- or three-bed flatlets, while the Guest Quarters can house 39 candidates in single rooms. As there are limitations on the number of places available, applications are subject to a selection process based on criteria such as the seniority of the candidate, his/her academic record, etc. Postgraduate candidates may obtain application forms and details regarding the cost of this type of accommodation directly from the Manager: Postgraduate Village.

Postgraduate candidates are not required to vacate the accommodation in the Postgraduate Village or Guest Quarters during the period from registration in February to the start of the University recess in December. Furthermore, master's and doctoral candidates who are required to remain on campus in order to continue their research during the official December recess may apply in writing for special permission to stay on campus to the Manager: Postgraduate Village (for attention: Student Housing Management-see below for contact details) motivating the reason(s) for the request.

This letter should be submitted by the beginning of October every year to serve at the relevant Management Committee meeting. It should be accompanied by a letter from the relevant candidate's Academic Manager confirming that the student will be required to stay on campus during the December recess period or will be required to return to the Village before the residences officially open at the start of the new academic year. Candidates will be notified in writing of the outcome of their application.

- Contact Details
- Centre for Teaching, Learning and Media
- Computer Services
- Department of Computer Science and Information Systems
- Department of Research Management

- Environment, Health, Safety and Cleaning Management
- Examination and Candidate Records
- Higher Education Access and Development Services
- Faculty Officers
- Arts
- Built Environment
- Business and Economic Sciences
- Computer Science
- Education
- Engineering
- Health Sciences
- Law
- Management
- Nursing Science
- Science
- Library
- Interlending Department
- Information Librarians per Faculty
- Arts
- Engineering
- Education
- Business and Economic Sciences and Law
- Sciences and Health Sciences
- Manager: Postgraduate Village
- Office for International Education
- Registrar
- Student Accounts
- Student Housing Management
- Writing Centre

DRAWING CONCLUSIONS

For any research project and any scientific discipline, drawing conclusions is the final, and most important, part of the process. Whichever reasoning processes and research methods were used, the final conclusion is critical, determining success or failure. If an otherwise excellent experiment is summarized by a weak conclusion, the results will not be taken seriously.

Success or failure is not a measure of whether a hypothesis is accepted or refuted, because both results still advance scientific knowledge. Failure is poor experimental design, or flaws in the reasoning processes, which invalidate the results. As long as the research process is robust and well designed, then the findings are sound, and the process of drawing conclusions begins. The key is to establish what the results mean. How are they applied to the world?

WHAT HAS BEEN LEARNED

Generally, a researcher will summarize what they believe has been learned from the research, and will try to assess the strength of the hypothesis. Even if the null hypothesis is accepted, a strong conclusion will analyse why the results were not as predicted. In observational research, with no hypothesis, the researcher will analyse the findings, and establish if any valuable new information has been uncovered.

GENERATING LEADS FOR FUTURE RESEARCH

However, very few experiments give clear-cut results, and most research uncovers more questions than answers. The researcher can use these to suggest interesting directions for further study. If, for example, the null hypothesis was accepted, there may still have been trends apparent within the results. These could form the basis of further study, or experimental refinement and redesign.

FLAWS IN THE RESEARCH PROCESS

The researcher will then evaluate any apparent problems with the experiment. This involves critically evaluating any weaknesses and errors in the design, which may have influenced the results. Even strict, 'true experimental,' designs have to make compromises, and the researcher must be thorough in pointing these out, justifying the methodology and reasoning.

For example, when drawing conclusions, the researcher may think that another causal effect influenced the results, and that this variable was not eliminated during the experimental process. A refined version of the experiment may help to achieve better results, if the new effect is included in the design process. In the global warming example, the researcher might establish that carbon dioxide emission alone cannot be responsible for global warming. They may decide that another effect is contributing, so propose that methane may also be a factor in global warming. A new study would incorporate methane into the model.

BENEFITS OF THE RESEARCH

The next stage is to evaluate the advantages and benefits of the research. In medicine and psychology, for example, the results may throw out a new way of treating a medical problem, so the advantages are obvious.

However, all well constructed research is useful, even if it is just adding to the fount of human knowledge. An accepted null hypothesis has an important meaning to science.

SUGGESTIONS BASED UPON THE CONCLUSIONS

The final stage is the researcher's recommendations based upon the results, depending upon the field of study. This area of the research process can be based around the researcher's personal opinion, and will integrate previous studies.

For example, a researcher into schizophrenia may recommend a more effective treatment. A physicist might postulate that our picture of the structure of the atom should be changed. A researcher could make suggestions for refinement of the experimental design, or highlight interesting areas for further study. This final piece of the paper is the most critical, and pulls together all of the findings. The area in a research paper that causes intense and heated debate amongst scientists is when drawing conclusions. It is critical in determining the direction take by the scientific community, but the researcher will have to justify their findings.

THE STRENGTH OF THE RESULTS

The key to drawing a valid conclusion is to ensure that the deductive and inductive processes are correctly used, and that all steps of the scientific method were followed. If your research had a robust design, questioning and scrutiny will be devoted to the experiment conclusion, rather than the methods.

THE PSYCHOLOGY OF AD EFFECTIVENESS IN DIFFERENT MEDIA ENVIRONMENTS

Cinema advertising is proven to be recalled with significantly more detail and deeper understanding than other media advertising. Behavioural psychologists prove the link between levels of emotional arousal (such as at the cinema) and the encoding of long-term memories on the brain

Executive Summary

We undertook a qualitative research study to observe the effects of media multi-tasking on advertising recall and depth of communication, and how this compares to a cinema environment (single-tasking).

Methodology

We used ethnographic research techniques to observe behaviour on a typical evening at two homes, and invited three groups (all groups were 20-30 year old ABC1s) to watch a new film at the cinema. None of the groups were aware they were going to be asked about advertising. We asked the groups to draw the ads they could remember after their evenings We consulted psychologists and analysed a popular ad model to help us understand the responses.

Key Findings

Observation

- Multi-tasking = Multi-media: a typical evening at home includes TV, Internet, mobiles, music, magazines, books, newspapers, etc.
- Five minutes was the maximum attention span on a typical evening – mostly much less
- There is a social etiquette at the cinema - NO talking, NO mobiles, look forward and pay attention

Psychologist's Perspective

- When multi-tasking, adults use selective attention to focus on what they're doing-they filter out the necessary information needed for each task
- When attention jumps from one thing to another, it is harder to recall information because the level of encoding (memory) is poorer
- However, it has been proven that memories associated with high levels of arousal are usually more easily recalled and are recalled in higher detail
- There's no distraction at the cinema which means ability to remember is improved
- Attention + emotion = an optimum level of arousal. This is key to increasing our ability to encode messages in the brain which in turn significantly increases recall

What does this mean for Advertisers

Drawings after a typical evening at home while multi-tasking show a thematic level of take-out (colours, characters and music are remembered but often unbranded). Drawings after a typical evening at the cinema display details (brand, facts, prices, dates, executional features) and meaning (interpretation and effect of the ad).

Why we Conducted this Study

Research frequently uncovers how busy our lives are. Juggling a number of commitments along with the plethora of new media platforms now available have not only changed the amount of messages we receive but also the way we consume media. Increasingly in today's environment we find several media channels being consumed at the same time. Media can switch from background to foreground and back again in seconds as audiences listen to the radio while surfing the Internet, dipping into magazines while watching TV, and texting throughout!

What effect does this new media environment have on advertising consumption? How can an advertiser be sure their commercial is achieving cut-through and if it is, how much detail or understanding is being taken out of the ad? Cinema's defining point of difference remains the audience's total focus and attention. So, exactly how much difference does it make to the 'take-out' of brand messages and understanding of an ad? What can behavioural psychology tell us about the link between emotional arousal and deeper encoding of messages in the brain? We conducted a piece of qualitative research with Work Research to better understand these issues. This report contains details of our methodology, the perspective of a leading psychologist, unique insight into what multi-tasking and single-tasking can mean for advertisers, and how this relates back to traditional ad models.

How we did it: Getting Closer to real life

We wanted to capture the reality of multi-tasking and single-tasking. Rather than ask people to remember what they *think* they typically do, we used the latest Big Brother style cameras to record their nights in, and out at the cinema to find out what they *actually* do. We used ethnographic techniques in order to observe people behaving as normally as possible.

Our research sample consisted of:

- One household of young men who had a typical evening in
- One household of young women who had a typical evening in
- Three groups of regular cinemagoers who were invited to the cinema to see a new film

Each household and the three groups were filmed for several hours and then asked to take part in an hour of group discussion. Unlike normal focus groups we didn't ask them to say very much.

Instead we used a new advertising recall methodology:

- We gave them a pack of coloured pencils and an art pad
- We asked them to draw all that they could remember from the adverts they had seen that evening

The drawings they produced speak volumes about the difference between multi-tasking and single-tasking media environments.

Key insights

Watching their Behaviour

- Multi-task = Multi-media - when multi-tasking they dip into magazines, newspapers, mobiles, Internet, TV, music, books
- The five minute attention span for multi-taskers - across the evening at home the longest period of attention was five minutes. Early in the evening when energy levels are higher, attention jumps around several times a minute
- A little less conversation! - men talk less, using media as prompts for comment. Women talk more, media fitting in around gossip
- The cinema experience has a set of defined behaviours which people conform to - "there's a kind of cinema etiquette". NO talking, NO mobiles, NO loud eating. Eyes forward, pay attention, focus, formulate opinions

Talking about their experience – what they said

- *MULTI-TASK*: "We talk non-stop, to be honest that's the main point of the night, I can't remember much else" "It's about crashing out, a few beers, your mind wanders"
- SINGLE-TASK: "You have paid and you want to get the most out of it so you really pay attention" "I just get there and settle in, it is just such an experience"
- *The Science Bit*: Psychology and neuro-science have lots to tell us about the effect of advertising and media experiences. We spoke to a number of psychologists to get their perspective.

Single-tasking, such as a visit to the cinema, often has its own etiquette which drives people to conform to appropriate behaviour, in this case being attentive, quiet, with no distractions. This means that people have focused attention. Their ability to remember what they see is improved because they are not being distracted by other things. In addition, they have invested in the experience by paying to be there, but more importantly choosing to be there.

Secondly, the high level of arousal the cinema experience creates (surround sound and visual) elicits emotion in the individual which increases their ability to recall what they have seen. This is because memories that are associated with high levels of emotion are more easily recalled and usually in more detail. Therefore, attention plus emotion increases ability to encode which in turn increases recall.

When multi-tasking, adults use selective attention to focus in on what they are doing. They filter out the necessary information needed for each task. When attention jumps from one thing to another it is harder to recall information because your level of encoding(memory) is poorer. You have less investment in what you are doing. You pay less attention and your ability to remember in these situations is impaired.

The Drawings: A model to Explain the Findings

We know that advertising is recalled in different ways.

- The most basic type of recall is thematic, where people typically remember basic visual featuressuch as colours, shapes or characters
- The next level of recall is more detailed, here people remember the facts contained in the commercial; the brand, prices, offers and executional details
- The most sophisticated level of recall requires a degree of interpretation and meaning, resulting in the understanding of the advert's deeper meanings and effect

Our research showed a distinct difference in recall between the multi-task and single-task drawings:

- Typically, multi-tasking results contained only limited thematic understanding
- Single-tasking offers both detailed recall and a communication of the advert's meaning

Conclusion

Cinema is arguably the sole single-task media experience, and such an experience provides a number of advantages to advertisers seeking stand out in today's multi-media, multi-tasking world.

- Cinema is the only unadulterated single-task medium left.
- We have observed a dramatic difference in people's behaviour when they are multi-tasking and single-tasking
- At home, the longest periods of attention were five minutes
- Cinema etiquette is both social (no talking) and personal (paying attention, thinking, absorbing)
- Ad recall is significantly more detailed following a single exposure in the cinema rather than in a multi-tasking environment
- Advertising has the ability to communicate deeper brand meaning to audiences who are at the cinema
- A high level of arousal cements experiences and messages in the brain.

Bibliography

Atsushi, I.: *The World of Nano-biomechanics: Mechanical Imaging and Measurement*, New Delhi: Oxford University Press, 2003.

Baird, D.: *Discovering the Nanoscale*, Hyderabad: Perspectives Publishers, 2001.

Bell, T.: *Societal Implications of Nanoscience and Nanotechnology*, NJ : IEEE Press, 2006.

Clayton, M.: *Nanoscale: Issues and Perspectives for the Nano Century*, New York: Harperbusiness, 2005.

Crawford, T.: *Nanotechnology Research*, New York: Harperbusines, 2008.

Drexler, K.: New *Nanotechnology Research*, New York : Morrow, 1999.

Drexler, L.: *Principles of Nanotechnology: Molecular-based Study of Condensed*, New York: John Wiley & sons, 1999.

Fujimasa, I.: *New Topics in Nanotechnology Research*, Oxford: Oxford University Press, 2005.

Hyungjun, K.: *Multiscale and Multiphysics Computational Frameworks for Nano*, London: Picador, 2005.

Kennedy, P.: *Choice: Publication of the Association of College and Research*, New York: Random House, 1999.

Mihail, C.: *Nanomechanics of Materials and Structures*, Boston: Kluwer Academic Publishers, 2005.

Nelson, M.: *The Potential of Nanotechnology for Molecular Manufacturing*, Philippines: Oxford University Press, 2008.

Newton, D.: *Sensors Based on Nanostructured Materials,* CT: Greenwood Press, 2006.

Philip, S.: *The Global Technology Revolution,* New Delhi: Institute of Objective Studies, 2006.

Ratner, *D.*: *The Chemistry of Nanomaterials: Synthesis, Properties,* NJ: Prentice Hall, 2006.

Regis, E.: *The Future of Tomorrow: How Technology, Medicine, Computers,* Boston: Little, Brown, 2004.

Smith, A.: *Research Directions in Distributed Parameter Systems,* New York: Random House, 1999.

Index

A

Adequate 15, 36, 47, 101, 115, 178, 201, 241
Affiliate 27
Afflicted 229
Allocated 42, 126, 169, 212
Anticipate 72, 75, 138
Anxiety 33
Appetite 89
Arbitrarily 48
Argued 35, 226, 233
Authentic 73, 74, 75

C

Citation 224, 235, 251, 253
Coefficients 20
Collaborative 56, 58, 68, 69, 70, 71
Compilation 73, 84, 88, 256
Conceptual 24, 38, 46, 58, 126, 150, 162, 258
Contagious 226
Contextural 56, 62, 63
Convergence 68, 72
Credibility 58
Criterion 47, 240
Critique 11, 57, 58, 59, 227, 250
Curriculum 63, 222, 249
Custodian 24

D

Deception 217
Deductions 72
Deductive 9, 10, 45, 60, 146, 226, 239, 240, 262
Demonstration 22, 157, 158, 199, 204
Desegregation 233
Dignity 34
Discerned 239
Disprove 22, 23, 138, 216, 217
Disputants 231

E

Edmonton 112
Empirical 1, 6, 39, 55, 60, 134, 148, 149, 150, 154, 241
Erroneously 28
Explicit 58, 59, 66, 74
Exploration 81, 89, 250
Extraneous 13, 14

F

Fatigue 226
Futuristic 224

H

Hermeneutics 60

Hydroponic 31, 32, 33

I

Illogical 91, 226, 227
Implementation 81, 82, 143, 214, 223, 254
Implicit 43, 57, 59, 146
Imputation 81, 82
Incompatible 39
Incrementalism 63
Incumbent 42, 46
Ineligible 101
Innovation 67, 151, 152, 154, 155
Insanity 226, 227, 228

L

Liberationist 63

M

Marginality 1, 2, 3, 7
Miraculously 226
Monetizing 27
Mystification 229

N

Naturalistic 10, 21, 69
Negotiated 58, 66

P

Paradigm 38, 39, 40, 56, 60, 61, 64, 253
Penetration 101
Potential 4, 13, 19, 33, 50, 58, 67, 69, 71, 72, 92, 98, 101, 104, 128, 130, 138, 198, 216, 254
Predict 2, 3, 5, 40, 47, 76, 101, 133, 136, 149, 153, 171, 178, 183, 184, 195, 205, 216, 225, 226, 234, 236, 237, 238, 240, 261
Predictability 5
Predictions 171, 184, 216, 225, 226, 234, 238, 240
Prestige 137, 177
Prostitute 224
Protocol 11, 91
Psychiatric 115, 123

R

Reflexivity 72
Regression 156, 195, 205
Relaxation 28
Reliability 46, 47, 48, 49, 92, 102, 151, 152, 154, 155, 239
Relying 55, 124, 154, 212, 219
Repulsive 225
Rigorous 36, 216, 217, 241

S

Sanctuary 225
Solicitation 224
Speculate 216
Stratum 44
Suspicion 23
Syllogism 9

T

Tabulation 159, 192, 252